Service Level Management in Emerging Environments

SCIENCES

Networks and Communications, Field Director – Guy Pujolle

Network Management and Control, Subject Head – Francine Krief

Service Level Management in Emerging Environments

Coordinated by
Nader Mbarek

iSTE

WILEY

First published 2020 in Great Britain and the United States by ISTE Ltd and John Wiley & Sons, Inc.

ISTE Ltd
27-37 St George's Road
London SW19 4EU
UK

www.iste.co.uk

John Wiley & Sons, Inc.
111 River Street
Hoboken, NJ 07030
USA

www.wiley.com

Library of Congress Control Number: 2020933380

British Library Cataloguing-in-Publication Data
A CIP record for this book is available from the British Library
ISBN 978-1-78945-002-6

ERC code:
PE7 Systems and Communication Engineering
PE7_1 Control engineering
PE7_8 Networks (communication networks, sensor networks, networks of robots, etc.)

MIX
Paper from
responsible sources
FSC
www.fsc.org
FSC® C013604

Contents

Preface

Nader MBAREK
LIB, University of Bourgogne Franche-Comté, Dijon, France

Communication technology has changed so much in our time that networks are omnipresent, with a variety of access equipment that allow people to constantly communicate with the network in order to produce or benefit from services provided by the network using ever-increasing and ever-more dynamic applications and operating environments. Networks have become an embedded part of daily life. This has been possible thanks to components that are constantly becoming smaller, cheaper and more powerful with higher processing speeds. Most important, perhaps, is that these components are able to form more and more connections between themselves using different technologies, thereby forming ambient networks on a large scale. In parallel with this quantitative explosion in communication networks, there is a complexification of technology. This rise in the number of networks and their complexity is accompanied by challenges in terms of management and control. Further, the increased competitiveness between service providers and the constant change in user needs requires genuine "dynamism" in service offerings, which is moving more toward Quality of Service (QoS), mobility and security as access networks and the technologies used develop further, while also trying to offer different service levels, thus responding to the needs of different types of applications. In this context, it becomes necessary to manage the service level demanded by the client and that the service provider must commit to offering. This service level is then used to efficiently parametrize the resources managed by the service provider based on customer needs.

It is from this point of view that different approaches for managing one or more service level components (QoS, security and mobility) are described in this book for different emerging environments such as the Internet of Things (IoT), the Cloud, smart grids, e-health, mesh networking, device to device (D2D), smart cities and green networking.

In Chapter 1, Ahmad Khalil, Nader Mbarek and Olivier Togni introduce the IoT environment through architectures proposed by different standardization bodies. They describe the importance of managing security and the protection of privacy in this kind of environment. Further, the authors highlight the needs and requirements of each layer in the IoT architecture in terms of QoS as well as in terms of the proposed QoS mechanisms that can respond to these requirements. Finally, a framework that uses a three-layer IoT architecture and a QoS-based access mechanism concerning the lowest level in this architecture are described.

In Chapter 2, Nader Mbarek describes the Cloud environment in terms of Cloud Computing, Cloud Networking and inter-Cloud. The author specifies the context and motives around guaranteeing service level in the Cloud in terms of QoS and security. A framework for guaranteeing the QoS in the Cloud Networking environment is introduced.

In Chapter 3, Samira Chouikhi, Leila Merghem-Boulahia and Moez Esseghir explore demand management as a service offered to consumers in a smart grid environment. They describe this environment and its essential components and also explain fundamental concepts of demand management. They focus on demand-side management, more specifically on energy consumption planning.

In Chapter 4, Mohamed-Aymen Chalouf introduces e-health systems, offering an overview of their architecture and characteristics. The author then describes the security and QoS challenges in this kind of environment and discusses several research papers and projects aimed at providing appropriate management solutions.

In Chapter 5, Hajer Bargaoui, Nader Mbarek and Olivier Togni introduce characteristics of wireless mesh networking, highlighting QoS management. The authors define routing concepts based on QoS in wireless mesh networks and describe Hybrid QoS Mesh Routing (HQMR) as an example of a routing protocol based on the defined QoS for this kind of environment.

In Chapter 6, Axel Moinet and Benoît Darties introduce the concept of blockchain through its basic characteristics and operating principle. The authors then describe how blockchain can fit into a model dedicated to the IoT in order to respond to authentication and trust challenges in this kind of environment. Blockchain Authentication and Trust Module (BATM) architecture is introduced as a solution that can respond to these challenges.

In Chapter 7, Chérifa Boucetta, Hassine Moungla and Hossam Afifi describe the emergence of D2D communications, focusing on the discovery of terminals in this kind of environment. It is essential, in this context, to optimize neighbor discovery in order to provide users with a high QoS while also ensuring the economizing of resources. The authors evaluate how an automatic learning strategy may contribute to a D2D environment.

In Chapter 8, Mohammed Salih Bendella and Badr Benmammar introduce not only green networking but also cognitive radio. They then study the impact of cognitive radio on green networking based on learning through reinforcement in order to extend the life of a network and reduce its energy footprint.

We hope that this book will help readers gain a greater understanding of the challenges and important issues related to service level management in emerging environments, as well as in their current works that focus on managing QoS, security and mobility in this kind of environment.

1

Service Level Management in the Internet of Things (IoT)

Ahmad KHALIL, Nader MBAREK and Olivier TOGNI
LIB, University of Bourgogne Franche-Comté, Dijon, France

1.1. Introduction

The Internet of Things (IoT) is now an integral part of our daily life. By 2020, there will be over 20 billion connected digital and electronic devices, which works out to about two devices per human being on Earth (Nordrum 2016). The IoT will thus have a significant impact on human life and will improve quality of life. The future growth of the IoT will lead to advanced use of technology in order to facilitate accomplishing daily human tasks. Consequently, improving corresponding services is an important challenge that must be faced in order to allow the expansion of this environment. In this context, it is expected that there will be a better user experience that will make up for the limitations experienced when using IoT services. User experience may translate to a service level that includes the expected Quality of Service (QoS) and also the expected level of security and privacy offered by the IoT environment. The objects connected to the IoT have certain restrictions in terms of memory, computing capacity and energy consumption. However, existing QoS security and privacy protection mechanisms do not take these constraints into consideration. Thus, it is primordial that we design and develop new QoS and security mechanisms or adapt and improve existing mechanisms in the context of the IoT.

In this chapter, we will first introduce, in section 1.2, definitions related to the IoT environment. We will then describe, in section 1.3, the architectures proposed

Service Level Management in Emerging Environments,
coordinated by Nader MBAREK. © ISTE Ltd 2020.

by different standardization bodies and the fields of application of the IoT. Section 1.4 introduces security management as well as the management of privacy protection in the IoT through the motivations, challenges and different security services that must be considered in this kind of environment. Section 1.5 describes QoS management by highlighting the needs and requirements of every layer of the IoT architecture in terms of QoS as well as the proposed QoS mechanisms that will respond to these. Section 1.6 defines our framework using a three-layer IoT architecture and a QoS-based access mechanism concerning the lowest level of this architecture. Finally, section 1.7 presents conclusions and perspectives related to service level management in an IoT environment.

1.2. IoT: definitions

Various standardization bodies have worked on the IoT in order to specify the definitions, architecture, recommendations and the fields of application for this new paradigm. The ITU–T (International Telecommunication Union–Telecommunication Standardization Sector) is a standardization body that works on the IoT environment and its different fields of applications through the SG20 work group. According to the ITU-T document Y.2060, the IoT is a ubiquitous network that is available everywhere, anytime and to anyone (ITU-T 2012). The IoT is a global infrastructure for the information society that makes it possible to offer advanced services by interconnecting objects using various communication technologies (Minerva *et al.* 2015). Further, in ISO/IEC (2015a), the ISO/IEC (International Organization for Standardization/International Electrotechnical Commission) provides a definition and specification for the vocabulary used within the IoT environment. According to this, the IoT is a network of physical objects that collect and transmit data. It is an infrastructure made up of interconnected objects, humans and information resources that make it possible to process data collected by the objects and then react as a consequence (ISO/IEC 2015a; International Electrotechnical Commission 2017b). According to the Internet Engineering Task Force (IETF), the general idea behind the IoT is to connect objects in order to provide contextual services across different technologies, thereby offering a service available anywhere and at any time (Minerva *et al.* 2015). The IETF considers the IoT to be a network of interconnected objects that can be addressed uniquely and that use standardized protocols for communication between the objects (Lee *et al.* 2012). Further, the IETF and ISO/IEC take different requirements into consideration in the IoT environment, such as auto-configuration, unique identification, interfaces standardization, connectivity, reliability and mobility.

We propose a definition that brings together information from all the definitions discussed above: the IoT is a global infrastructure that interconnects objects (which are identified uniquely) and humans to offer advanced, autonomous services via smart interfaces. It must be noted that the unique identification allows the identity of the objects to be verified and enables data processing based on the source of the data.

1.3. IoT: an overview

1.3.1. *IoT architectures*

Standardization bodies and research projects have introduced various architectures for the IoT. In the following sections, we describe two examples of proposed architectures for the IoT.

1.3.1.1. *The ITU-T reference model*

The ITU-T proposed a reference model for the IoT that is based on multiple layers (ITU-T 2012). This reference model (see Figure 1.1) defines four horizontal layers (application, support, network and devices) and two vertical layers (management and security). The application layer includes IoT applications and services. Next, the Service Support and Application Support Layer defines the capacities for generic support that are common to all applications, such as data processing and storage. This also includes specific support capacities that respond to the needs of a particular application. The Network Layer offers two services. First there are network capacities, which ensure that connectivity, mobility management, authentication, authorization and accounting functions are all monitored. Second, there are transport capacities that control the routing of data coming from applications or information from the monitoring and management of the environment. The Device Layer defines the capacities of each connected object and device as well as the capacities of the communication gateways.

The vertical layers define the generic management capacities (management of objects, network, traffic and congestion) as well as the security capacities (authorization, authentication, integrity and privacy protection). Further, these vertical layers introduce specific capacities that are dependent on the type of IoT application being considered.

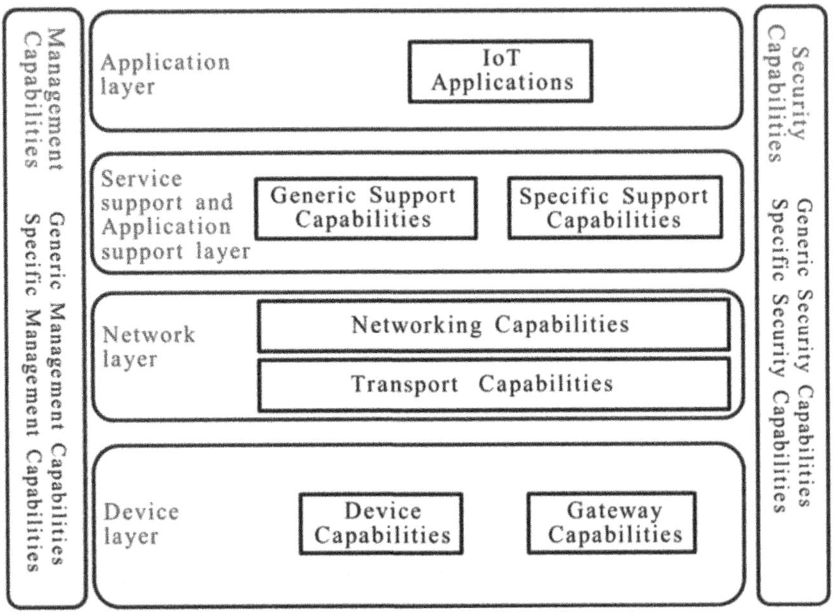

Figure 1.1. *The ITU-T architecture of the Internet of Things (ITU-T 2012). For a color version of this figure, see www.iste.co.uk/mbarek/service.zip*

1.3.1.2. *The IIC architecture*

The Industrial Internet Consortium is a consortium of several well-known industrial groups in the IT world, such as IBM, HUAWEI and Intel. Through the Industrial Internet Reference Architecture report, this consortium puts forth a system architecture that is applicable to the IoT. This three-tiered architecture is based on three vertical layers or three levels (see Figure 1.2) (Lin *et al.* 2015).

The Edge Tier corresponds to all the nodes that collect data from proximity networks. This layer makes it possible to implement all control functions. Then comes the Platform Tier, which receives, processes and transmits control commands to the Edge Tier. This layer also enables the processing, analysis and running of operations on data collected from objects, before transmitting them in the opposite direction, toward the Enterprise Tier. The Enterprise Tier takes decisions and carries out the role of an interface with the end-user. It thus includes applications that allow control commands to be generated and to be sent to the Platform Tier. The different layers in this architecture are interconnected via access networks and service networks.

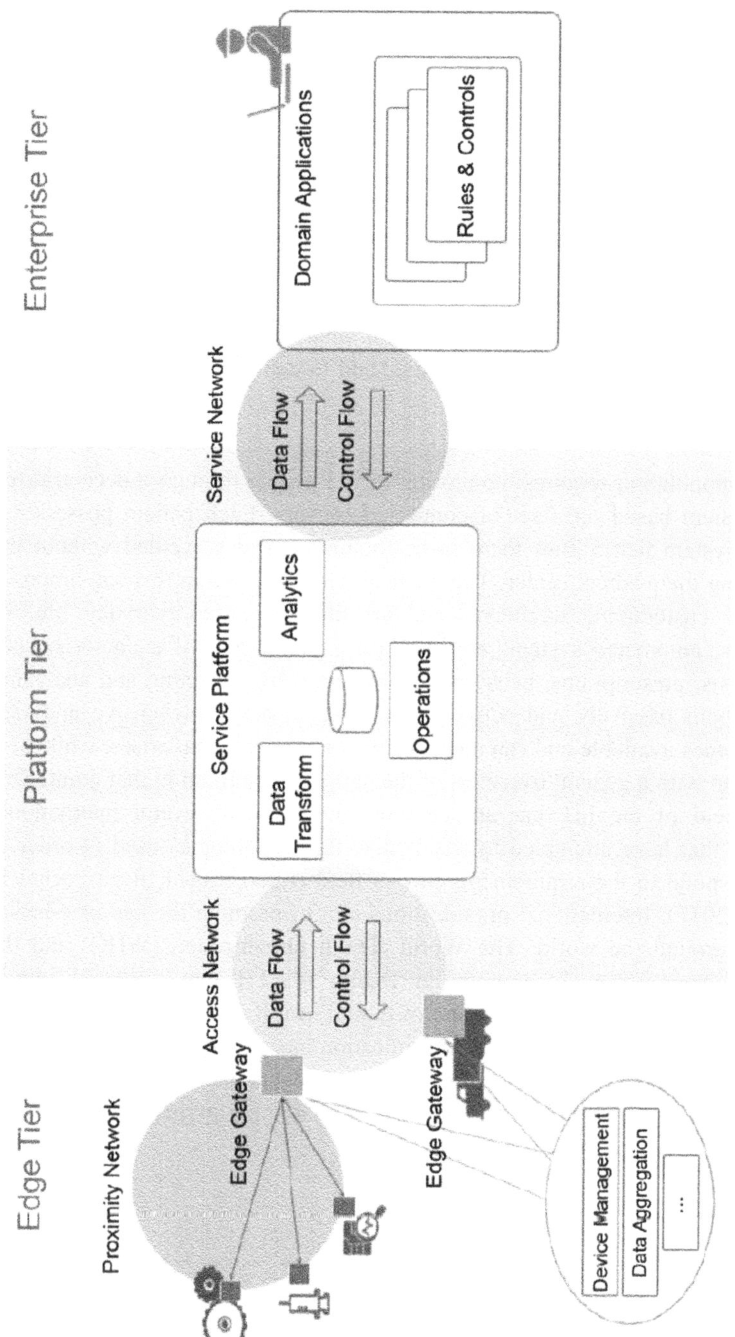

Figure 1.2. *Architecture of the Internet of Things (Lin et al. 2015).*
For a color version of this figure, see www.iste.co.uk/mbarek/service.zip

1.3.2. *Application fields of the IoT*

The IoT improves the quality of life in different areas of daily life. Examples include the field of health, smart cities, vehicular networks and so on. The ISO/IEC focuses on the standardization of the underlying technology used in different fields of application of the IoT. Working Group 9 under Technical Committee 1 (JTC 1/WG9) of ISO/IEC studies the normalization of Big Data technologies in IoT domains (International Electrotechnical Commission 2017). Further, various providers offer solutions for IoT service offerings and implementation in different fields of application. For example, the Kaa[1] project offers a range of features allowing us to create advanced applications for smart devices, to flexibly manage ecosystems and their peripherals, to orchestrate end-to-end data processing, etc.

1.3.2.1. *E-health*

An aging population requires monitoring of old people through a decentralized healthcare system based on a set of connected sensors. Each patient possesses a surveillance system that allows them to be monitored and surveilled without the need of visiting the medical center. The medical data collected in this way improve healthcare by customizing treatments and creating an easier everyday life for patients. Thus, automated systems can perform a major part of a doctor's work (tests, diagnosis, prescriptions, behavior modification) by collecting and analyzing patient data both passively and actively. As a result, a comprehensive and rich database becomes available and can alert doctors to any need that arises, while also providing them with a general overview of the patient's health up to that point. This application field of the IoT has attracted the attention of several international organizations that have attempted to standardize the technologies used in order to effectively respond to the requirements of this field (International Electrotechnical Commission 2017). International organizations aim to promote the use of e-health technologies around the world. The World Health Organization (WHO) and the Program for Appropriate Technology in Health (PATH) have entered into a partnership to accelerate the development of digital health around the world (World Health Organization 2018). This field of application has attracted a large number of industrial organizations that try to offer different products that would be useful for e-health. Indeed, Ericsson and its partners offer portable prototypes for the field of e-health with long battery lives (Ericsson 2018).

1 Available at: https://www.kaaproject.org (Accessed November 02, 2018).

1.3.2.2. *Smart cities*

In the context of smart buildings, the IoT offers management systems that can automate control functions through intelligent devices while providing real-time data analysis. Because of the data collected from the IoT objects, it is possible to anticipate needs and preferences in terms of lighting, heating or ventilation. Similarly, this information may relate to security systems, electricity meters or the removal of waste and sewage. Reactivity with respect to these data that are collected through IoT objects allow buildings to adapt to changes and to implement the required modifications in real time, for increased efficiency, and to reduce operational costs. Putting in place smart buildings makes it possible to create smart cities by integrating other IoT devices into different services offered by the city such as transport, water and air quality. Various standardization organizations have focused on this field through working groups that study the standardization of the technologies used. For example, ISO and IEC, through the technical subcommittee JTC1/SC25, have established norms for microprocessor systems and interconnection supports associated with equipment for commercial and residential environments (International Electrotechnical Commission 2018). Further, the IEC white paper (2017a) describes smart buildings and smart cities as being fields of application of the IoT. This paper specifies how to orchestrate the infrastructure required for smart and sustainable cities. Even as these groups establish norms, Nokia (2018) has launched several services and technologies on the market to manage video surveillance, the sensor networks for the IoT, parking and the environment within a smart city. In this context, infrastructure has been proposed with applications that make it possible to transform any contemporary city into a smart city.

1.3.2.3. *Vehicular networks*

All types of transportation systems may benefit from the advantages offered by the IoT. IoT solutions promise to make transportation systems more intelligent and better performing by improving safety, the efficiency of their journeys, the maintenance of vehicles and by offering more strategic traffic-management (Alcatel Lucent Enterprise 2018). Communication systems between vehicles and infrastructure (V2I) and communication systems between vehicles (V2V) enhance safety, efficiency and the performance of public and private transport. They also contribute to reducing congestion and improving space management. Drivers of connected cars can benefit from a large number of services such as navigation, real-time traffic and parking information, as well as the integration of smartphones with the dashboard and portable devices (International Electrotechnical Commission 2017). The revolutionizing of the transportation world by applying the IoT in the field of vehicular networks has been possible because of the use of sensor networks and applications for parking management, traffic management, etc. For example, smart roads use sensors to determine the number of cars in each lane and then manage traffic lights based on this information so as to minimize congestion. The effectiveness of this field of application of the IoT can be seen in the implementation of this transformative technology in different projects. For example, the ParkDC project, implemented by the Washington D.C. transport department, uses a surveillance system based on the IoT to alert drivers to parking spots that are available and to calculate the appropriate parking charges based on real-time demand (Njit 2018).

1.4. Security management and privacy protection in the IoT

1.4.1. *Motivations and challenges*

The security of information systems is made up of all technical, organizational, legal and human resources required to prevent the unauthorized use, misuse, modification or hijacking of the information system. At present, security is a major challenge in the information world and the goal of security in this context is to maintain the trust of the users and the consistency of the entire information system. Several norms have arisen around concepts related to security, for example the X800 recommendation by ITU-T (1991), which emphasizes the role played by different security services and their applicability.

The IoT is characterized by an environment that is subject to constraints across several levels, which makes it difficult to adopt security mechanisms that were designed for conventional systems. An IoT environment includes objects with low memory resources and limited computational power. Further, the techniques normally used in conventional networks were designed for systems that contained powerful microprocessors and had high storage capacities (Hanna 2015). Existing security techniques must thus be adapted. Further, the large number of objects in an IoT environment makes it a difficult and onerous task to adapt existing security algorithms. For example, methods and algorithms for identification and controlling access to objects become more and more complex as the number of objects in the environment keeps increasing.

Before a device or a user can access IoT services, mutual authentication and authorization between the device/user and the IoT system must be established in accordance with predefined security policies. Security policies must be drawn up with great precision in order to comprehensively cover all possible use cases and must also follow standardized models in order to respond to the requirements of the IoT. It is, therefore, important to standardize security policies for the IoT environment. Further, access to data or services must be entirely transparent, traceable and reproducible. This results in an enormous volume of trace files created in the IoT environment given the large number of connected objects. Thus, the mechanisms to optimize traceability must be designed for the context of the IoT. In this kind of an IoT environment, a variety of operating systems with different architectures are available for IoT objects. We can cite here, among others, the example of Google's Android Things (formerly called Brillo) (Google's Internet of Things Solutions 2018), Huawei's LiteOS (2018) and Windows 10's IoT Core (2018). This diversity can make it even more difficult to standardize security mechanisms and measures.

As concerns user privacy, data can be collected in IoT systems without involving the users. In this context, this data feedback must be secured and the user's privacy must be ensured during the collection, transmission, aggregation, storage, extraction and processing of the data. In order to meet these requirements, the appropriate mechanisms for data confidentiality, data authentication and data integrity must be included within the IoT, while respecting the needs of this kind of environment (ITU-T 2012).

A number of international organizations have worked on concepts related to security and privacy in the IoT, either by offering appropriate security mechanisms or by offering methodologies that can be applied across the layers of their IoT architectures. We thus have the ITU-T Y.2060 recommendation (ITU-T, 2012) that aims to secure the IoT environment by starting with an analysis of the threats that are specific to the IoT application. Then, specific security services and mechanisms will be supported at every layer of the IoT architecture to ensure global security within this environment. In terms of the application layer of the ITU-T reference model, different security services will be considered, such as authorization, authentication, privacy and integrity of application data, and also the protection of privacy. As concerns the network layer, the security services include authentication, confidentiality of the application data and the signaling data (configurations and commands) and the protection of the integrity of the network management techniques. For the lowest level of ITU-T IoT architecture, namely the device layer, the main services and mechanisms offered to guarantee security are authentication, authorization, validation of the device integrity, access control, confidentiality of data and the protection of integrity. Following the recommendation (ITU-T, 2014), several specific security abilities must be considered in the IoT environment: the ability to ensure secured communications to guarantee the confidentiality and integrity of the data during transmission and during storage. Further, the recommendation specifies an ability to provide a secure service that guarantees that fraudulent services will be forbidden and an ability for authentication and mutual authorization between objects and users in accordance with predefined policies to guarantee the security of information access. They are closely tied to the specific needs of IoT applications and depend on their field of application. Recommendation Y.2060 (ITU-T 2012) also emphasizes the need for security functions and mechanisms to be supported by IoT gateways interconnecting the different components of the different layers of the IoT architecture specified by ITU-T. In the following section, we will describe the different security services that must be considered in the IoT environment.

1.4.2. *Security services in the IoT environment*

In order to ensure security in the IoT environment, various security services must be provided by applying mechanisms that are specific and adapted to the characteristics of this kind of environment.

1.4.2.1. *Identification and authentication in the IoT*

1.4.2.1.1. Definition

Identification refers to establishing the identity of the user of a service. It is based on the principle of each user being individually assigned an identifier. Authentication follows identification and enables the user to prove their identity. The user should use an authenticator or a secret code, which only they know. Authentication does not give the right of access. It is the access control that guarantees this privilege if authentication has been successful (ITU-T 1991). Authentication mechanisms can offer several advantages to the IoT environment. Thus, through the identification and authentication mechanisms, the IoT environment takes into account robust devices that are able to reduce the risk of intrusion and avoid violations (Li 2017).

Further, conventional identification and authentication methods must be adapted to meet the requirements of the IoT environment in terms of scalability, the large number of entities, etc. Several organizations use digital certificates based on public key infrastructure (PKI) for device identification and authentication operations (Allerin 2018). However, certain adaptations must be carried out in order to consider this solution in the IoT context. First of all, the PKI infrastructure must be able to effectively support the process of issuing digital certificates in large numbers and at high speeds. For example, a cloud-based PKI is a more economical and realistic method for the scale required by an IoT environment. Second, digital certificates have a limited lifespan, which means they have an expiry date. In the context of an IoT environment, some use cases may require short-term certificates while many others require certificates with a longer lifetime. A certificate with a longer lifetime is required when a device needs to be authenticated on the basis of a long-term certificate. IoT project managers must then carefully determine the lifespan required for digital certificates and determine the associated advantages and disadvantages. On the other hand, the certificates, which are considered critical elements with their own life cycle, must be managed in an efficient manner. The manual tracking of these certificates is not feasible in an IoT environment. Thus, PKI must be associated with certificate management providers along with a scalable platform. These platforms must then be capable of managing specific IoT use cases (Allerin 2018).

1.4.2.1.2. Research projects

Various research studies and projects have dealt with identification and authentication security services. BUTLER (uBiquitous, secUre inTernet-of-things with Location and contExt-awaReness) (CORDIS 2018), a European project funded by FP7 (October 2011–October 2014), studied the mechanisms of identification and authentication in the IoT environment. This project proposed a mechanism for managing the ownership of objects by the users. In this case, users possess connected objects. A user (the owner of an object) has an account with the Trust Manager, which is implemented on an authorization server. The user connects to the authorization server and registers a new resource (a new connected object). The resource must have a unique identifier (generally a URL) and identification information (resource security credentials). The user must then configure the resource with the resource security credentials and, thus, the identity of the user who possesses the object may be verified. Similarly, BUTLER offers a mechanism that makes it possible to identify objects to gateways using digital certificates that are managed by authorization servers (Sottile *et al.* 2014). There is also academic research that has studied identification and authentication in the IoT. According to the work described in Li (2017), the author highlights the importance of proposing an authentication protocol that makes it possible to relieve nodes (which are constrained in terms of their storage and computing capabilities) of the management of authentication and authorization.

1.4.2.2. *Access control in the IoT*

1.4.2.2.1. Definition

Access control makes it possible to fight against the unauthorized use of a resource. In order to implement this control, a list of entities authorized to access a resource with their access authorizations is defined in accordance with a security policy. This security service is offered to implement different types of access to resources (reading, writing, modification, information deletion and task execution). Access control is based on one or more elements, using an information database that is maintained by authorization centers or the entity itself and this may take the form of an access control list or a hierarchical or distributed matrix. These databases include authentication information (passwords, security tags, etc.) (ITU-T 1991). Two entities are used for access control in the IoT: the data holders (users of IoT services) and objects (data collectors) that send data or receive commands. These two entities must be mutually authenticated (Balte *et al.* 2015).

1.4.2.2.2. Research projects

Several European research projects have studied the adaptation of access control mechanisms for the IoT environment. ARMOUR (2018) is a European project funded by H2020 (February 2016–February 2018) that addressed some of the challenges surrounding security and trust in the IoT. The work carried out in the framework of this project makes it possible to define a set of components that interact with each other to authorize or block secure data queries in an IoT environment. ARMOUR defines several entities in this environment in order to do this. First, we have the Policy Decision Point (PDP), which is a component that includes the access policies and, by evaluating the access control policies, can authorize or deny authorization to an IoT device (sensor) to carry out an action on a resource (data registration server). For example, a "PERMIT" decision from the PDP allows the Capability Manager (the server communicating with the PDP) to generate and send a token to the sensor to publish the data on the IoT platform. The data publication server (Pub/Sub Server) saves the data and thus allows the data query to be updated and executed if the sensor token received by the Capability Manager allows this action (ARMOUR 2016).

SMARTIE (Secure and SMArter ciTIes data management) (Pokric *et al.* 2015) is another European project funded by FP7 (September 2013–December 2016) focused on access control in the IoT. SMARTIE's goal was to develop new mechanisms to establish trust and security in the different IoT layers. The results of the project indicate that Attribute-Based Access Control (ABAC) is an appropriate solution to specify finer access control policies. In ABAC, the identity of an IoT service user is no longer limited to a single attribute but is based on multiple attributes (i.e. user ID, role, etc.) that make up this identity. This is why ABAC provides substantial improvements in authorization and access control within the IoT. ABAC-based solutions make it possible to overcome the disadvantages of centralized access control solutions. Each query requires two steps: an authorization check (identity control and authentication) and, consequently, an access control decision (authorization or prohibition). For each access request, the IoT service user is authenticated with the domain and the access authorization request is obtained for the user. The user's access authorization request is signed by a trusted domain authority. Thus, the user may send the query to the IoT devices that verify the signature. If the signature is successfully verified, the required information is sent to the user (SMARTIE 2014a, 2014b; Pokric *et al.* 2015).

1.4.2.3. *Confidentiality in the IoT*

1.4.2.3.1. Definition

The confidentiality service offers protection against non-authorized entities analyzing traffic and against data flows being divulged. Data encryption is the most appropriate mechanism to ensure this security service. Encryption can be carried out using a symmetric system (with a secret key) or an asymmetric system (public key). Symmetric encryption involves knowing the secret key that allows encryption and decryption. For asymmetric encryption, the knowledge of the public encryption key by all entities does not imply knowledge of the private key for decryption. Apart from encryption mechanisms, there must be a key management mechanism in order to exchange keys between the communicating entities (ITU-T 1991).

In an IoT environment, there are several points that must be taken into consideration when using the confidentiality service, especially during the key exchange process for encryption. First of all, extensibility is an important characteristic that must be considered, as there is a high number of connected objects. In fact, the number of entities that can be involved in the key exchange process may be limited by using conventional systems. Second, new entities may be involved after the initial key exchange. Thus, new objects may be integrated into the IoT environment after the initiation of the services. Scalability is another important characteristic that must be taken into consideration. Indeed, when new entities are involved in the key exchange process in the IoT, the volume of cryptographic data to be stored on the objects becomes greater, while IoT objects are subject to restrictions in terms of data storage and processing abilities (Abdemeziem 2016).

1.4.2.3.2. Research projects

One of the challenges in implementing an encryption system for a connected object in the IoT environment is the availability of appropriate software libraries that respect the constraints governing IoT objects in terms of memory, computation ability and energy consumption. In this context, certain research projects have been carried out to address this problem, which still poses a challenge and requires more advanced studies that are better adapted to the needs of the IoT in order to provide optimal security services. An example of an existing library that can be used in an IoT environment is the "AVR-Crypto-Lib" (Cantora 2013), which provides special implementations that respect the limited resources of microcontrollers. This library offers symmetric key encryption such as AES, RC5, RC6 and DES. Another library, "Relic-Toolkit" (2018), offers a large variety of asymmetric encryption algorithms such as RSA and Rabin crypto system. "Relic-Toolkit" is used in the TinyPBC project implemented on the TOSSIM simulator (2018) on the TinyOS operating system. The libraries we have just described provide, among other things, a confidentiality service in an IoT environment, which allows secure communications, so that unauthorized access to the content of the data is prohibited and that content is protected during its transfer between two entities in the IoT environment.

European research projects have also focused on data confidentiality on the IoT. The SMARTIE project (Pokric *et al.* 2015), for example, uses CP-ABE (Ciphertext Policy Attribute-Based Encryption), a technique that allows the IoT user to decrypt the message from objects with a secret key if the policy attributes match the attributes of the key. CP-ABE makes it possible to encrypt data for a group of users, instead of encrypting it individually, in accordance with access policies. This technique links access control and encryption and is used when data from an object must be received by several users of that IoT service. Data are thus encrypted only once (SMARTIE 2014a; Pokric *et al.* 2015). The European project BUTLER (CORDIS 2018) is focused on the protection of the communication channel in the IoT. This channel is vulnerable because of its wireless feature and information dissemination. BUTLER proposes improvements to security standards used in IoT communication technologies. For ZigBee, it offers a security system based on the use of symmetric keys to complement and enhance the security features provided by the ZigBee standard, which uses two mandatory keys and one optional key. The Master Key and the Network Key are mandatory, while the Link Key is optional. The Master Key is used in the initialization phase and implemented at the nodes through an out-of-band channel. The Network Key guarantees the security of the network layer and is shared by all nodes. It is derived from the Master Key. The optional Link Key is derived from the Master key and guarantees the security of the link between two peers at the application level. In this context, the BUTLER project put in place mechanisms to manage the deployment, maintenance and revocation of the Master key. It also proposed implementing an additional symmetric key (called the Global Key) at the node, at the time of manufacture. This key is used by the Medium Access Control (MAC) layer and is shared by all nodes. The Global Key guarantees security for the lower layers. The Network Key provided by the ZigBee standard will thus be used as a Group Key, which will be shared between the nodes and managed by the ZigBee Network layer. This makes it possible to securely address a group of nodes sharing a common feature. As a result, objects communicating via ZigBee will be guaranteed greater security as well as additional security when using the optional Link Key (Sottile *et al.* 2014).

1.4.2.4. *Integrity in the IoT*

1.4.2.4.1. Definition

Integrity is a security service that covers two significant concepts in the IoT: the integrity of data and the integrity of objects. The integrity of data aims to ensure that the data exchanged in an IoT environment is not modified or destroyed in an unauthorized manner during transfer. This is necessary in order to provide a reliable service and ensure that the information collected and commands received by the objects are legitimate. Verifying the integrity of data involves two processes, one involving the sender and the other the recipient. The entity that is transmitting the data adds verification information (like the Block Check Character or a cryptographic check value such as a hash value) based on the data transmitted. The recipient generates the same verification information based on the data received and compares this information with the information received in order to determine whether or not the data were modified during transmission in the IoT environment (ITU-T 1991).

The integrity of objects is necessary as the nodes in the IoT may be deployed in an unreliable environment and may be physically attacked to modify the software codes in the objects, for example. This second integrity service in the IoT enables the detection and prevention of any modification to the operating system and the configuration of the objects. The integrity of objects also makes it possible to lock and eliminate non-compliant devices. To implement this type of integrity, a digital fingerprint for the object in question is used to compare data effectively available on the object with the data that should be available.

1.4.2.4.2. Research projects

Various European research projects have studied the security service ensuring both types of integrity, that of data as well as of objects, in an IoT environment. SMARTIE, for instance, takes into account several architectures for the implementation of the integrity service in an IoT environment. It uses Linux's kernel integrity measurement architecture (Pokric *et al.* 2015) to verify the integrity of objects. It additionally takes the support of integrity verification mechanisms present on smart cards, while taking inspiration from the Integrity Measurement Architecture (IMA). SMARTIE thus offers a node-attestation component that makes it possible to verify the integrity of the node by testing the hashing for the list of software and files that have been executed on that node. The node attestation component consists of a Remote Attestation mechanism between IoT objects and the remote central unit that is responsible for measuring the integrity of the objects. Remote attestation allows the remote party – the gateway or server responsible for verifying the integrity of the objects – to inspect the state of a device or an IoT object at any given moment. The remote party may request the hashing of the list of software or files and is able to verify whether the records provided by the device have been falsified by comparing the hashing received with the hashing that was calculated. The node-attestation component developed in SMARTIE makes it possible to provide a practical solution, which is a compromise between the hardware solution and the software-based approaches by using the IMA module and the architecture for integrity measurement that is present in the Linux kernel (SMARTIE 2014a; Pokric *et al.* 2015). The IMA module measures the integrity of the binary code before the kernel proceeds to loading the code into memory to be run. The measurement result is recorded and sent to the IMASC service (Integrity Management Architecture using a Smart Card). The IMASC system transmits the result to the smart card, where it is timestamped and signed so that there can be no subsequent manipulation of the entry. In addition, the smart card preserves a record with the hash value. For remote attestation, the verifying party can inspect the state of a remote device at any time by requesting the hashing and verifying the signatures. During a remote attestation request, the IMASC service interacts with the smart card and with the remote party in order to provide the proof of attestation. Further, various libraries have been designed for IoT objects in order to carry out the hash functions. For example, we have "Cryptosuite" (Knight 2010), which is a library for Arduino that supports different hashing algorithms such as SHA-1, SHA-256, HMAC-SHA-1 and HMAC-SHA-256.

1.4.2.5. *Non-repudiation in the IoT*

1.4.2.5.1. Definition

The non-repudiation service ensures that one party cannot deny its involvement in exchanges. This service can take one or two of the forms described below: the first form is non-repudiation with proof of origin, where the recipient receives proof of the origin of the data. This proof may be a digital signature using asymmetric encryption applied to the result of the hashing of the data exchanged. The second form is non-repudiation with proof of the data delivery, where the sender receives this proof in the form of an acknowledgment, for example (ITU-T 1991). The non-repudiation security service is necessary in the IoT to provide proof of data transmission through objects and also as a proof of the dispatch of any order by users of IoT services. This may fit into the framework of an audit that will allow the tracking and recording in trace files of all events that took place in an IoT environment.

1.4.2.5.2. Research projects

The first form of non-repudiation (i.e. with proof of origin) is based on mechanisms that are used to guarantee integrity, such as the data signature.

Consequently, the adaptation of non-repudiation mechanisms to an IoT environment may borrow from the adaptation of integrity services in the IoT. Non-repudiation was guaranteed in SMARTIE by the implementation of a signature for the list of software and operating systems of objects in order to verify the identity of the hashing issuer (see section 1.4.2.4.2).

1.4.2.6. *Availability in the IoT*

1.4.2.6.1. Definition

Availability refers to the possibility of on-demand access and use of resources by an authorized entity, following authentication and access control. Availability is a security service. Indeed, a service that becomes unavailable after a Denial of Service (DoS)-type attack, for instance, is an unsecured service and may be compromised at any time (Mosenia and Jha 2017). Availability in the IoT is essential to provide an Internet-enabled environment that is completely operational. In the context of the IoT, this service integrates both the availability of devices (that is, objects and gateways), allowing for uninterrupted data collection, and also the availability of IoT services offered to users. This second type of availability is determined by the configuration of the IoT environment, resulting in the need to make appropriate choices for management and administration protocols and for protection against DoS attacks (DDoS: Distributed DoS). Services offered in the IoT must be constantly available while taking into account the critical nature of some of these services (Mosenia and Jha 2017). In this context, availability round-the-clock, 365 days a year, is required for critical IoT services, such as certain applications in the field of e-health.

1.4.2.6.2. Research projects

In accordance with recommendations made by the GSMA (Global System for Mobile Communications Association), a commercial organization that represents the interest of 800 mobile operators around the world, availability was brought to the fore, while indicating that nodes must be capable of continuous communication with each other, with users and with back-end services (GSMA 2016). The European iCore project ((Internet Connected Objects for Reconfigurable Ecosystem) (Menoret 2012), funded by FP7 (October 2011–October 2014), defines the security requirements that must be taken into consideration in a framework concerning the IoT. The framework, called the “Open Cognitive framework” takes into account three levels: Virtual Objects (VO), which offers a virtual representation of objects; Composite Virtual Objects (CVO), which represents the fusions of several VOs; and the user level. This framework makes it possible to ensure that objects are re-used in these services, thus allowing a certain redundancy in order to improve availability. Similarly, the fusion of some VO through the CVOs ensures better availability. A semantic description of object capabilities allows this reutilization. The iCore project also puts forth recommendations to be respected in different practical use cases in the IoT. Thus, the report (Menore 2012) emphasizes the idea of providing mechanisms to protect information infrastructure against DoS threats and to implement the mechanisms required to support the recovery of service after a failure.

The research described in Nagara *et al.* (2017) specifies a portable DoS testing tool that is based on a software for IoT devices. This DoS test must be carried out at the design and development stage of the product. The tool consists of an attacking entity and a monitor. The attacker carries out a DoS attack on the target device (that is, the IoT object) and transmits information on the traffic to the monitor. In this context, the "Mirai" attack was used and targeted the devices using Linux to transform them into "bots" that could be remotely controlled and used for a large-scale network attack. The tool effectively verifies whether the IoT devices are resistant or vulnerable to DoS-type attacks.

1.4.3. *Privacy protection and trust in the IoT*

1.4.3.1. *Privacy*

Protecting privacy in the IoT requires specific considerations to protect information related to a person's privacy from being shared in this kind of environment. Data transmitted by a single object may not generate confidentiality problems that could compromise an individual's privacy. However, when fragmented data originating from several different objects is re-assembled, compiled and analyzed, it may generate sensitive information that requires appropriate protection.

As already seen in section 1.3.2, the IoT operates in different fields of application through which personal data of the users is collected. In fact, IoT service consumers risk divulging private information, little by little, without realizing it as they are unaware of the nature of the data collected and how it is used in this kind of environment. Current approaches to data protection in IoT are based chiefly on encryption or on access control to the collected data. Nonetheless, threats to privacy in the IoT may not be covered by the mechanisms that these solutions offer. For example, data processing may be outsourced, creating the risk of data being sold for marketing or other purposes to third parties (Sicari *et al.* 2015).

1.4.3.2. *Trust*

Trust is a complex concept influenced by many measurable and unmeasurable properties. Trust is closely related to the security of systems and users and is a necessary condition. However, trust is related not only to security but to several other factors, such as the QoS provided by the system, its reliability, its availability and, therefore, the services offered by the system and so on. Another important concept related to trust is the protection of privacy. A trust system must protect the privacy of its users so as to gain their trust. Trust, security and privacy protection are crucial issues in the emerging domain of information technologies such as the IoT (Yan *et al.* 2014).

Trust is managed through several processes from data collection to the provision of customer service. Trust management in the IoT thus provides an efficient means of evaluating trust relations between IoT entities and helps them in taking decisions about communicating and collaborating with each other. To guarantee this trust, data detection and collection must be reliable in the IoT. Special attention must, therefore, be paid to properties of trust in this kind of environment. These properties include the sensitivity, accuracy, security, reliability, and persistence of the object, as well as the effectiveness of data collection. This collection generates an enormous volume of data that must be carefully processed and analyzed, maintaining trust in terms of reliability, protection of privacy and accuracy. Further, the data must be securely transmitted and communicated in an IoT environment. An important challenge to face in meeting this objective is managing encryption keys in the IoT, as data confidentiality is common to security services, privacy protection and trust management. Moreover, the measures needed to act against attacks that could impact all levels of the IoT environment must be taken into consideration and we must ensure that the system is robust against all types of attacks in order for users to be able to sufficiently trust their IoT environment. Finally, users expect scalable and efficient identity management. Identity management concerns all layers of the IoT architecture, starting from the object all the way up to the user of services hosted in the Cloud. Identity management must respect the confidentiality of the service user's identity in order to respect their privacy. The context of the IoT service is likely to influence identity management strategies. For example, a critical IoT service in the field of e-health requires finer and more specific identity management (Yan *et al.* 2014).

1.4.3.3. *Regulations*

European States and the European Union have put forth various regulations for the protection of privacy. One of the most important of these is Directive 95/46/EC of the European Parliament, which emphasizes the protection of individuals with respect to the processing of personal data and the free circulation of such data (Official Journal L 281 1995). There is also the General Data Protection Regulation (GDPR), which is the European Union regulation that serves as the reference text with respect to the protection of private data. It came into effect on May 25, 2018 and brought about a significant change in the processing of user data. This regulation must be respected when offering services in the IoT and includes various requirements (Official Journal of the European Union 2016):

– process personal data equitably, lawfully and transparently;

– only collect and preserve personal data that are truly necessary and destroy these data after use;

– ensure that the data held are accurate and updated;

– ensure that you are ready to manage the increased rights of individuals;

– designate a data protection officer;

– develop and maintain a register of processing activities;

– share personal data responsibly.

1.4.3.4. *Research projects*

Several research projects have been carried out on trust and privacy in the IoT. For example, there is the report from the AIOTI (Alliance of Internet of Things Innovation) Workshop on Security and Privacy in IoT, which outlines the key security and privacy requirements for different application fields of the IoT. These requirements can be summarized as the user being able to monitor the data, the transparency and control of the user interface, default encryption, data insulation, continuous monitoring, etc. Further, this report highlights the importance of applying additional mechanisms such as minimizing data collection and the need for accountability in the misuse of collected personal data (AIOTI 2017).

There is also TCG (Trusted Computing Group 2018), which is a group formed by AMD, Hewlett-Packard, IBM, Intel and Microsoft, which aims to implement the concepts of "Trusted Computing" in personal computers. In this context and through the report (Hanna 2015), the TCG's IoT subgroup defined a trust system as a system designed to be predictable even under stress. This same report specifies that to build a trusted IoT system, it is necessary to set up a hardware Root of Trust (RoT), use encryption during storage, add automation of security and protect legacy systems. The RoT can generate random numbers, store and use long-term keys and verify the integrity of the system in order to reduce risks and provide the system with strong protection. In this context, the Trusted Platform Module (TPM) is an open and interoperable ISO/IEC standard (2015b) that can specify a hardware RoT. The technical specification of this standard has been written by the TCG group. TPM provides security features such as authentication, encryption and attestation (guaranteeing the security of software or hardware to a third party). At present, TPM is integrated into billions of connected objects. Hardware Storage Encryption is a component that provides the encryption service. Hardware Storage Encryption uses Self-Encrypting Drive to provide continuous encryption with no impact on system performance. This component makes it possible to protect against physical attacks, loss or theft by instantly erasing data and cipher suites. Security automation makes it possible to automatically process tasks related to security operations. This automation is brought in for all phases of system security (incident detection, incident analysis, incident response) (Hanna 2015). This security automation is standardized across different standards such as IEEE 802.1AR (IEEE Standards Association 2018) and TAXII (Trusted Automated eXchange of Indicator Information) (Davidson and Schmidt 2014).

Finally, the research work carried out by Dong *et al.* (2011) presents a trust management model in the IoT called TRM-IoT. This model helps defend against malicious node attacks by identifying them and taking the appropriate decisions. To perform this task, it is necessary to measure the trust of objects and this is done by taking into account complex aspects such as credibility assessment of nodes, for example. Fuzzy logic theory is then used to determine trust by taking into account different input parameters: the end-to-end packet transmission ratio, energy consumption and the packet delivery ratio.

1.5. QoS management for IoT services

1.5.1. *Motivations and challenges*

The IoT is a combination of different technologies and heterogeneous networks providing various types of services. Consequently, several types of data and flows with different or even contradictory constraints can coexist in the same IoT environment. From this observation follows the need for integrating QoS to satisfy the needs of each data type (Gubbi 2013). QoS is a critical element in the IoT, given that the large number of connected objects leads to a considerable increase in the volume of data in this type of environment. This increase may create congestion that affects the performance of the IoT system. In this context, QoS mechanisms optimize performance by identifying different traffic and applying the differentiated treatment. Performance management also helps reduce costs and improve scalability in the IoT environment (Bhaddurgatte and Kumar 2015).

QoS models used in traditional computer networks such as IntServ (Braden *et al.*, 1994) and DiffServ (Babiarz *et al.*, 2006) are difficult to implement in the IoT. This difficulty arises due to the constraints in terms of memory and computation capacities that IoT objects have as well as the wide scope of IoT and the random deployment of these objects. Thus, new and adapted QoS models must be offered to provide IoT services with a guarantee of connectivity, performance and, generally, of QoS (Premila *et al.* 2015).

Various international organizations have emphasized the importance of guaranteeing QoS in the IoT. ITU-T studied the importance of integrating QoS in the IoT environment in Y.2066 (ITU-T 2014), stating that prioritizing IoT services is an essential requirement. In addition, Y.2066 indicates that the prioritization feature makes it possible to satisfy the different service requirements for various IoT user groups. In other words, differentiated services should be supported, so that the IoT can provide different Service Level Agreements (SLAs). On the other hand, LinkLabs, an American company developing technologies for computer networks, indicates that the integration of QoS into IoT allows better management of the capabilities and resources in this type of environment to provide reliable and optimized infrastructure for connectivity of IoT objects. According to LinkLabs, in order to offer secure and predictable IoT services, QoS mechanisms will help manage delays, jitter, bandwidth and packet loss by categorizing traffic and recording the limits of communication channels (Ray 2016). The following section examines the requirements in terms of guaranteeing QoS as well as the responses provided within the different layers of the IoT architecture.

1.5.2. *Guaranteeing QoS in IoT*

1.5.2.1. *IoT device layer*

1.5.2.1.1. QoS requirements

The device layer of the IoT architecture includes connected objects and gateways that manage these objects. The QoS at this level must, therefore, respond to requirements with respect to objects and gateways.

An essential requirement for ensuring QoS in the device layer is the differentiation and prioritization of flows. The IoT brings together a varied group of applications of different types, belonging to different fields. Consequently, data from different applications do not have the same priority and different flows must be categorized based on their criticality and a certain prioritization must be applied through various appropriate QoS mechanisms. Based on the same concept of heterogeneity of applications, the QoS mechanisms applied must match the characteristics of these different applications. It is thus important to categorize IoT applications based on specific characteristics so as to offer the QoS mechanisms defined in the context of a homogeneous set of applications. Each set of applications will therefore have mechanisms that are well adapted to their requirements.

The performance management of the system, through the optimization of resource use to offer the best performance, is an important mechanism that we have to propose to ensure QoS at the level of this layer of the IoT architecture. The performance of the IoT environment must be monitored in order to meet the needs of clients using the IoT applications and to provide a service that fulfills the corresponding requirements. This monitoring consists of collecting precise information on the system and therefore the parameters for measuring the performance of the IoT environment must be well chosen. This choice makes it possible to specify parameters that are specific to the IoT, which must be quantified using the appropriate metrics.

Different protocols and communication technologies are used for information exchange between objects and gateways within the device layer in the IoT architecture. Therefore, the QoS mechanisms related to the device layer depend on these technologies. Some of these technologies have common bases, which makes it possible to generalize some of the QoS mechanisms used in this layer.

1.5.2.1.2. Research projects

Various projects and research studies have proposed the integration of QoS into the device layer of IoT architecture using mechanisms that have been adapted for this environment.

OpenIoT, a European project funded by FP7 (December 2011–December 2014), specifies different parameters for QoS in the IoT (Serrano 2014). These parameters are varied and can be divided into several categories. First of all, we have the general parameters such as the quality of the sensors that determine the accuracy and sensitivity of the measurements, and also the energy consumption. In addition, OpenIoT also takes into consideration the amount of data produced by a physical sensor, the coverage of the network of objects, and the trust correlated to the sensor quality. On the other hand, various parameters related to the network of objects in the device layer must be taken into consideration, for instance, lifetime, delay and jitter. Different metrics are considered to quantify each QoS parameter specified in the OpenIoT project. For example, the parameter related to energy consumption is quantified using battery level. Finally, this project makes it possible to apply QoS using a QoS Manager that can collect information on different parameters, which will allow the manager to take decisions in order to improve the quality delivered to users (Serrano 2014).

The research work carried out in Nef *et al.* (2012) concerning QoS guarantee in the IoT suggests that the different IoT applications be classified according to three service models, i.e. Open Service Model, Supple Service Model and Complete

Service Model. Each category will correspond to a physical topology for the implementation of the sensors. The Open Service Model refers to interactive, non-real-time and non-critical applications; the Supple Service Model corresponds to interactive, Soft Real Time and critical applications; the Complete Service Model corresponds to interactive, Hard Real Time and critical applications. The authors then classified applications belonging to different domains according to these three models. In addition, this work has matched the proposed service models to physical topologies (star topology and random topology) at the device layer to meet the needs of each model. Indeed, applications belonging to the "Complete" model must be provided using a physical star topology for shorter delays. However, applications belonging to the "Open" model must be provided through a random physical topology for better energy consumption.

Other research has focused on prioritizing access to the shared communication channel between IoT objects to ensure there is a differentiation of traffic when sending data to gateways. The research work carried out in Sayarifah *et al.* (2015) defines different queues and a scheduler in an object in order to prioritize the processing of priority flows. Other research has attempted to adapt Slotted Carrier Sense Multiple Access/Collision Avoidance (CSMA/CA) access algorithms to ensure that QoS is guaranteed. The research conducted in Feng *et al.* (2013) adapted the IEEE 802.15.4 standard, and in particular the CSMA/CA algorithm, by integrating a QoS mechanism. This new access method, called SDA-CSMA/CA, is based on the specification of different values for CSMA/CA variables (Contention Window [CW] and Maximum Backoff Exponent [BE]) to create different priorities depending on the types of traffic. Thus, three priority levels are considered with SDA-CSMA/CA. In this context, smaller CWs and narrower BE intervals (minimum BE and maximum BE) are assigned to high priority traffic. As a result, important messages have a greater chance of accessing the channel with reduced waiting times.

1.5.2.2. *IoT network layer*

1.5.2.2.1. QoS requirements

The network layer of the IoT architecture includes all the network functionalities such as routing, transfer and path management (path selection and recovery) via a multipath infrastructure. This layer acts as a network infrastructure interconnecting the object network with the Cloud infrastructure.

The integration of QoS in this layer will enable the processing of the large volume of requests and information passing from the device layer to the support layer of the IoT architecture. This processing must be differentiated; in other words, it must prioritize the requests according to their importance. As a result, the QoS requirements for the network layer of the IoT architecture correspond to the QoS requirements of a network infrastructure while also adapting these requirements to the characteristics of the IoT environment. Various mechanisms, such as the prioritizing of requests and the choice of paths based on a QoS context, may be useful in meeting the requirements of the IoT.

1.5.2.2.2. Research projects

F-Interop Platform, a European project funded by H2020 (November 2015–November 2018), has developed test tools for QoS and Quality of Experience (QoE), based on SDN/NFV (Software Defined Networking/Network Functions Virtualization) that can be used in the IoT (Ziegler 2015). These technologies can replace conventional technologies to guarantee QoS in the network layer of the IoT architecture. SDN technology consists of separating monitoring functions from data routing functions. The SDN controller (the component centralizing the decision making and management of the network formed by network equipment which exchanges data with each other) can impose policies on flows to provide different levels of QoS. NFV technology, on the other hand, makes it possible to provide modules to apply the various policies for the deployment of agents and to emulate network conditions while collecting measurements related to QoS parameters (Ziegler 2015). The collection of QoS statistics (delay, jitter, etc.) may be carried out through the F-Interop solution using a passive or active method. The passive method relies on the collection of QoS statistics from SDN switches on which counters are implemented to measure the throughput or loss of packets, for example. The active method relies on the use of NFV agents to measure QoS parameters in real time. The collection of information relates to data regarding packet loss, delay and jitter (Ziegler 2015).

1.5.2.3. *The support and application layer in the IoT*

1.5.2.3.1. QoS requirements

The support layer of the IoT architecture includes Cloud services as well as computing and storage capacities. In addition, the application layer includes IoT applications for processing data and carrying out calculations for useful purposes.

It is difficult to guarantee QoS in the Cloud hosting IoT applications and this constitutes an emerging discipline that poses several research challenges. This is due to the shortage of standardized end-to-end approaches for QoS guarantee in this type of environment. In addition, there are a variety of constraints and QoS parameters specific to each Cloud service. Indeed, the QoS requirements in the Cloud are varied and depend on the service provided by the Cloud. These services are Infrastructure as a Service (IaaS), Platform as a Service (PaaS) and Software as a Service (SaaS). The Cloud can also provide network services called Network as a Service (NaaS) that may make up the infrastructure connecting the device layer to the support layer offering IaaS, PaaS or SaaS type services in an IoT environment.

QoS requirements in a Cloud environment, and therefore in the support layer and the application layer of the IoT architecture, may relate to the availability of the service, the need to provide mechanisms that can process different types of data flows in a differentiated manner, the allocation of resources according to the needs of data flows or standardization of SLAs (Hamze 2015).

1.5.2.3.2. Research projects

Various projects and research works have studied the integration of Cloud technology in the IoT as well as the guarantee of QoS in this context.

The European BigClouT project, funded by H2020 (July 2016–June 2019), aims to use the three major technological tools of the moment, namely IoT, Cloud and Big Data, in order to increase the efficiency of urban, economic and natural infrastructure shared by growing populations (Grugen 2018). Another research study carried out in Khodkari *et al.* (2017) considered that the integration of the Cloud in the IoT environment makes it possible to guarantee a QoS level for applications because of the significant resources of the Cloud. Indeed, the Cloud is able to manage diversified data and supply IoT applications with resources in a dynamic way to guarantee performance while lowering costs. In this context, the authors have proposed different functions for measuring QoS attributes (that is, response time, availability, throughput, etc.). On the other hand, the authors in Chowdhury *et al.* (2018) study the Cloud environment and also IoT. In this context, they highlight different concepts concerning the maintenance of good QoS, the different parameters on which QoS depends (classifying them according to IoT layers), the different problems encountered in maintaining these parameters and the possible solutions for overcoming issues in the Cloud and IoT. This work presents the various parameters to be taken into consideration for the IoT architecture layers, such as traffic load, service deadlines, service time, service precision and service priority. According to this research, QoS parameters in the IoT environment must also consider characteristics of IoT applications hosted in the Cloud.

1.5.2.4. *Cross-layer approach in IoT*

1.5.2.4.1. QoS requirements

Having defined the various QoS requirements inherent to each IoT layer and the mechanisms that can enable QoS in each layer, we must specify the requirements and mechanisms to provide end-to-end QoS across the different layers of the IoT architecture. This end-to-end QoS allows clients to perceive the service level provided. The client (IoT application user) evaluates the overall level of end-to-end QoS and does not differentiate between the different levels into which the QoS is divided.

Different research studies have focused on end-to-end QoS in the IoT and specify the different needs relating to this type of environment. First of all, the IoT environment must be capable of prioritizing certain flows and services based on their requirements. IoT users who have paid for privileged services must be able to benefit from these. In addition, the deployment of the IoT, starting from the device layer up to the hosting of the application at the support layer, must be carried out in an optimal way to take into account the capabilities required from the different components. The gateways must be able to process the volume of data transmitted by the objects while the Cloud service must be able to provide the necessary computing and storage capabilities. Similarly, all components of the IoT environment must be capable of adapting themselves in real-time to meet the needs of different user flows. The QoS requirements for different layers of the architecture must then be translated such that they are comprehensible to the other layers. This translation allows different IoT layers to communicate with each other in order to meet QoS requirements. Consequently, guaranteeing end-to-end QoS in the IoT first requires breaking down QoS needs according to the different layers of the IoT architecture and then finding mechanisms to apply in a complementary manner between these layers.

1.5.2.4.2. Research projects

Various projects and research studies have been carried out with the aim of studying the guarantee of QoS in the IoT environment based on a transversal approach involving the different layers of the IoT architecture. We present below a non-exhaustive list of proposals produced by these studies.

SymbIoTe, a European project funded by H2020 (January 2016–January 2019) (Zarko *et al.* 2015), aims to offer a system providing IoT services if the availability and QoS potentially associated with this can be guaranteed by the underlying infrastructure. The IoT system must prioritize privileged users over basic users if there are simultaneous attempts to access certain IoT services. This priority must be ensured across all IoT layers in order to guarantee end-to-end QoS. In addition, the system must allow access control to IoT services in accordance with local and global legislation (security of access to critical data is a necessary requirement requested by several governments and international organizations), the current overload, security issues, etc. This access control must be provided across all IoT layers (Zarko *et al.* 2015).

In addition, several research projects have studied the transversal approach to guaranteeing QoS in the IoT environment. The research work presented in Duan *et al.* (2011), for example, is based on the principle that two adjacent layers in the IoT architecture communicate with each other to provide QoS. The upper layers send a QoS query to the layer that is directly beneath, which translates the request into a number of QoS parameters. This interlayer communication is carried out using dedicated interfaces with clearly specified parameters. The QoS for the support and application layer in the IoT architecture is directly perceived by the clients and is based on parameters such as time of service, precision, load and priority. The network layer QoS depends on the network type and is based on parameters like bandwidth, delay, jitter, and packet loss ratio, etc. Finally, QoS for the device layer is based on parameters of sampling, coverage, synchronization and mobility. It is useful to differentiate services in order to identify different service levels and attribute these to clearly defined IoT applications. In addition, each level corresponds to a number of specific QoS requirements. The research work conducted in Duan *et al.* (2011) organized applications into four classes depending on the type of task to be performed. The four categories of applications specified are as follows: Guaranteed Service, Guaranteed Service/Differentiated Service, Differentiated Service, and Best Effort. Therefore, IoT control applications require Guaranteed Service, IoT queries require Guaranteed and Differentiated services, IoT real-time surveillance applications require Differentiated Services, while non-real-time surveillance applications require Best Effort services.

1.6. QBAIoT: QoS-based access method for IoT environments

In order to guarantee a service level within the IoT, in Khalil *et al.* (2017) we propose a framework based on an architecture that enables QoS guarantee through the establishment of different types of SLA.

1.6.1. *Service level guarantee in the IoT*

1.6.1.1. *QoS-based architecture for the IoT*

We propose the architecture illustrated in Figure 1.3 in order to allow an IoT service provider (IoT-SP) to provide a service to a client (IoT-C) in accordance with a service level agreement (iSLA:IoT-SLA).

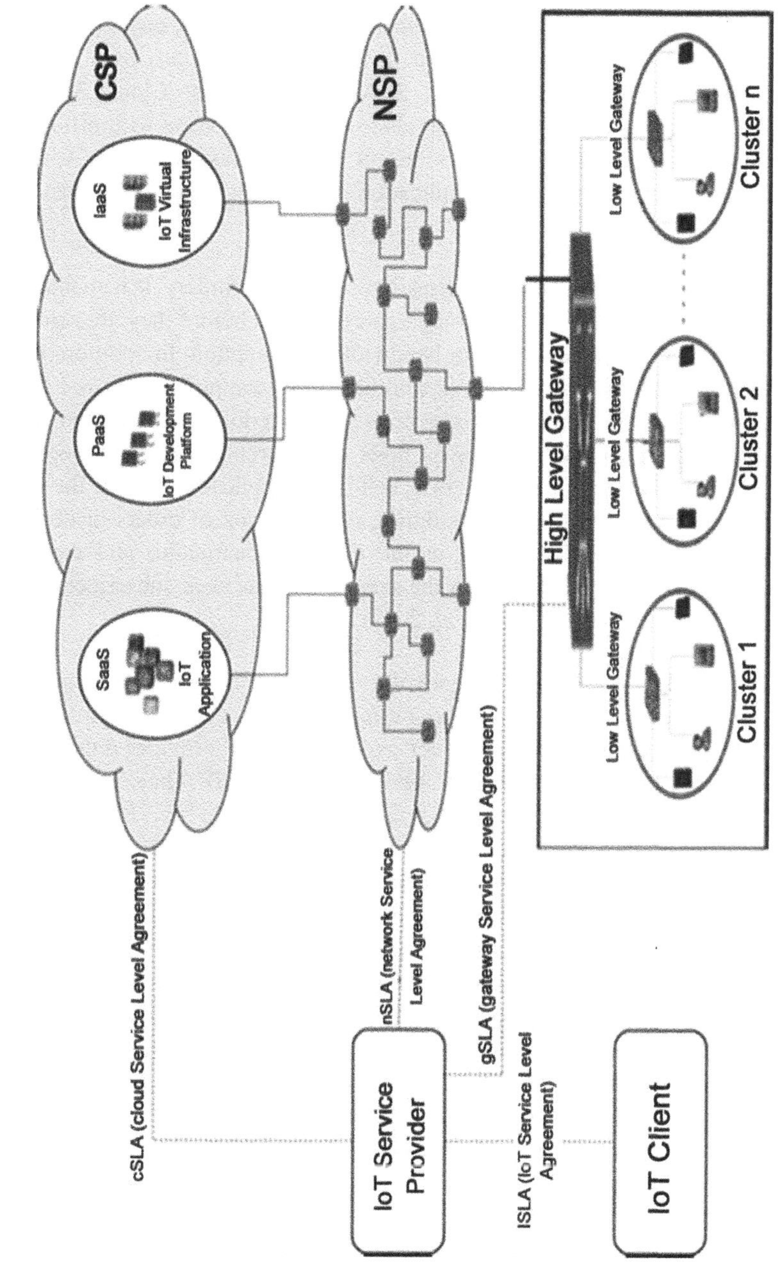

Figure 1.3. *Architecture of the Internet of Things (Khalil et al. 2018)*

The proposed architecture is based on three layers (sensing, network and Cloud) and different entities, each playing a specific role. IoT-C requests a service from the IoT-SP. The requested service uses an infrastructure of objects owned by the customer or the IoT-SP. In addition, the IoT-SP may have its own Cloud infrastructure or it subscribes to specific SLAs with different Cloud Service Providers (CSP). On the other hand, IoT-SP also subscribes to SLAs with different network service providers (NSP) to interconnect object infrastructures to Cloud resources. The IoT service requires an application part and an object infrastructure part.

The object infrastructure retrieves information or executes orders. It is managed by a High-Level Gateway (HL-Gw) that aggregates the data before they are sent to the Cloud in order to guarantee minimum bandwidth consumption. In addition, this HL-Gw offers local data processing capabilities (fog computing). Low-Level Gateway (LL-Gw) on the other hand classifies the data for differentiation and ensures the application of QoS at the lowest layer of the proposed IoT architecture (sensing layer or device layer, according to ITU-T). The application part of the IoT service enables the monitoring of objects during the execution of orders or during data processing. It is generally hosted in the Cloud infrastructure and can be provided in different forms depending on the service that has been subscribed with the CSP (IaaS, SaaS and PaaS) (Khalil *et al.* 2017).

An SLA between the provider and the customer allows the expected characteristics of the IoT service to be defined. In the context of our IoT architecture, the SLA must include customer expectations and provider guarantees with respect to the three layers (sensing, network and Cloud). Thus, our global IoT-SLA (iSLA), established between the IoT-SP and the IoT-C, is based on different sub-SLAs established between the IoT-SP and CSP (called cSLA: cloud SLA), and the NSP (called nSLA: network SLA). In addition, an internal SLA is established between the IoT-SP and the HL-Gw (called gSLA: gateway SLA). The cSLA includes performance and QoS parameters depending on the type of subscribed Cloud service. The performance parameters of an IaaS service differ from those of the PaaS and SaaS services. Thus, a SaaS service response time may be sufficient for measuring performance, while for an IaaS service, the cSLA client is offered an infrastructure and, consequently, performance must be measured through other parameters, such as computing capacity and storage capacity. The nSLA is based on conventional network QoS parameters such as bandwidth, latency, jitter, packet loss ratio and availability. The gSLA describes the processing and storage capacities used by the IoT service at the HL-Gw. It also includes the characteristics of LL-Gw and those of the object infrastructure used for offering the IoT service (Khalil *et al.* 2017).

1.6.1.2. *Service level agreement in the IoT*

The IoT-SP establishes a service level contract called iSLA with the IoT-C using a set of subcontracts such as cSLA and nSLA, implemented with various NSPs and CSPs. The global iSLA type agreement makes it possible to identify the IoT service using the "Service Identification" attribute. At the client's end, this attribute is based on a list of IP addresses used by the IoT-C. At the supplier's end, an application identifier, the IP address of the application interface, the port number as well as the identifiers of the HL-Gw and LL-Gw are used. In addition, the iSLA-type agreement includes an attribute for "IoT Service Performance Parameters" that describes the characteristics of the IoT service using three subattributes: "IoT Sensing Characteristics", "IoT Application Characteristics" and "IoT QoS Parameters". Thus, the "IoT Sensing Characteristics" section includes the characteristics of the gateways (HL-Gw and LL-Gw) and of the group of IoT objects (i.e. inter- and intragateway communication technology, number of objects connected by LL-Gw, number of LL-Gws per HL-Gw, etc.). The "IoT Application Characteristics" section includes parameters related to cSLA such as the application identifier, the maximum number of users, the average number of requests per second per user, the support time and the storage parameters. Finally, the "IoT QoS Parameters" section includes qualitative and quantitative QoS parameters. The qualitative parameters define the type of application by specifying the QoS class of the data generated. We can cite the example of the category of services called Real-Time Mission Critical (RTMC), those called Real-Time Non Mission Critical (RTNMC), the category of Streaming services and the category of Non-Real-Time (NRT) services. The quantitative parameters depend on the type of service covered by the iSLA and include the end-to-end delay, availability, the lifespan of the objects and the quality of the data (standard deviation, frequency of detection and error ratio of the data, etc.). The iSLA-type agreement also includes a "Business Parameters" attribute specifying the cost of the IoT service and the methodology for managing violations and resulting penalties through thresholds and monitoring intervals for each quantitative QoS parameter.

1.6.2. *The QBAIoT process in the IoT*

In order to guarantee the QoS parameters negotiated in the iSLA service level agreement, we propose a new QoS mechanism for the lowest layer of the IoT architecture, namely the sensing layer. We thus specify a wireless access method called QBAIoT (QoS Based Access method for IoT) (Khalil *et al.* 2018). This method enables the communication between the IoT objects and the LL-Gw, while adapting the superframe structure of the IEEE 802.15.4 standard and the slotted CSMA/CA mechanism. The adaptation allows the differentiation of traffic in order to meet the QoS requirements of each class.

The QBAIoT method takes into account up to four QoS classes as defined in the iSLA's qualitative parameters (RTMC, RTNMC, Streaming and NRT). Real-time traffic (RTMC and RTNMC) is very sensitive to delay, Streaming traffic is sensitive to jitter (that is, variation of delays), while NRT traffic is a QoS class that has no QoS constraints. To adapt the structure of the superframe, QBAIoT replaces the single Contention Access Period (CAP), common to all objects, and also replaces the non-contention period (CFP: Contention Free Period) of the classic IEEE 802.15.4 superframe by novel QoS CAPs. Each QoS CAP is specific to a QoS class. Thus, the adapted QBAIoT superframe can contain up to four QoS CAPs. The number of QoS CAPs configured in a LL-Gw depends on the number of QoS classes defined in the iSLA for this gateway. In addition, the inactive part of the classic superframe is eliminated in the context of QBAIoT. This deletion leads to shorter timeframes and enhances the performance of traffic in the Real-Time category. On the other hand, the number of slots available in the superframe is fixed at 16 and the duration of each slot depends on the superframe duration (SD). The SD is calculated based on the value of the Superframe Order (SO). The Beacon Order (BO) is used to calculate the interval for sending beacon frames (Beacon Interval [BI]). In QBAIoT, BO and SO are always equivalent as the inactive period is deleted.

During each QoS CAP in the QBAIoT superframe, only the objects that generate traffic belonging to the corresponding QoS class can compete for access to shared support. Each QoS CAP is assigned a number of slots among the 16 available in the superframe. The configuration of the slots and of the BO/SO values depends on the number of existing QoS classes in the IoT environment under consideration and, in particular, the number of real-time QoS classes. If only one class exists, the classic IEEE 802.15.4 superframe structure will be used, but with no CFP period and no inactive period, as only one QoS CAP exists. In addition, in this case, the values of the BO and SO are set to 14 to minimize the number of beacons sent in the IoT environment. When several QoS classes exist, the BO and SO are initialized to a value of 2 if there is at least one real-time QoS class (RTMC or RTNMC), and to a value of 3 if there is no real-time QoS class. The QBAIoT method is implemented on the LL-Gw (acting as coordinator) as well as on the IoT objects of the sensing layer. Figure 1.4 shows a comparison between the structure of the standard IEEE 802.15.4 and QBAIoT superframes.

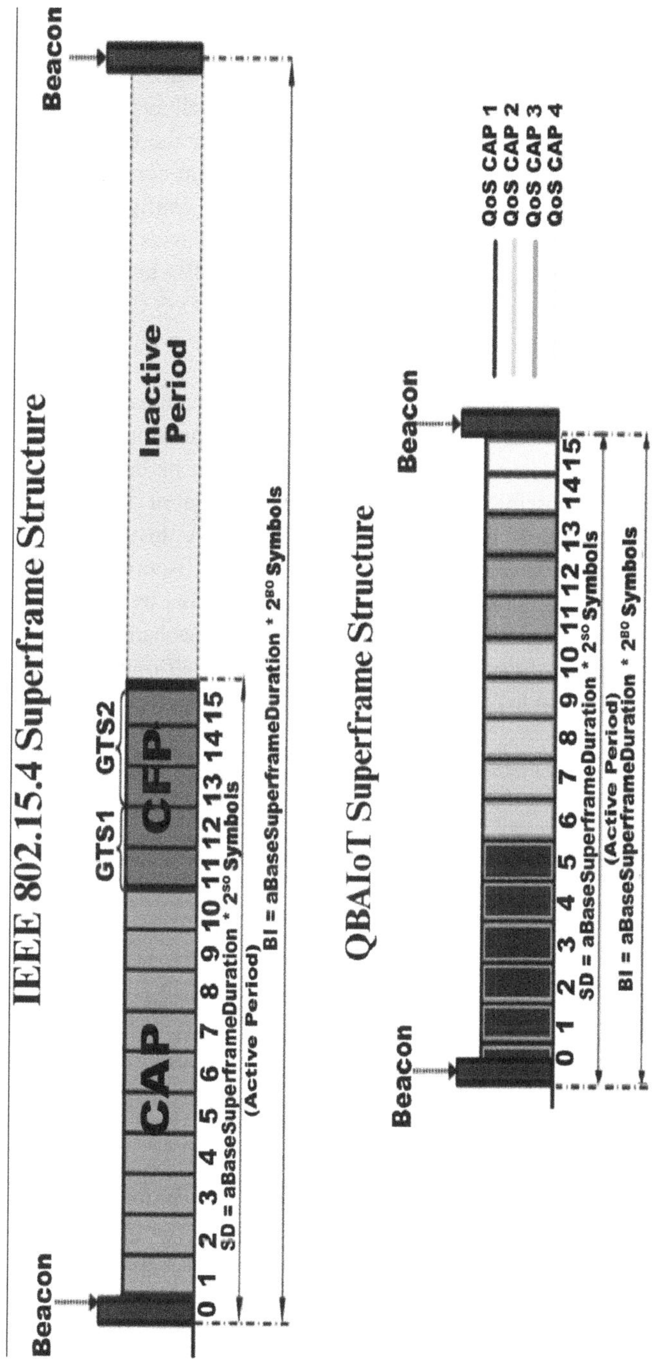

Figure 1.4. *Comparison of the structure of the IEEE 802.15.4 and the QBAIoT superframes. For a color version of this figure, see www.iste.co.uk/mbarek/service.zip*

1.6.2.1. *QBAIoT gateways*

For each BI, the LL-Gw that implements QBAIoT sends a beacon frame containing the information about the BO and SO values as well as the values for the first and last slot of each QoS CAP. These values are used by the IoT objects in the sensing layer to determine how long they are allowed to be in contention to access the channel. The QBAIoT gateway chooses an initial configuration for the superframe according to the number and types of QoS classes that exist in its environment. Following the dispatch of a beacon frame, the QBAIoT gateway begins to receive data from different objects during the corresponding QoS CAPs. The schema for the QBAIoT algorithm deployed for the gateway is illustrated in Figure 1.5.

A QBAIoT gateway also has self-management functionalities. Indeed, a self-configuring function allows the gateway to adapt the configuration of the different QoS CAPs based on the existing number of QoS classes in its environment. This function also makes it possible to apply a reconfiguration of these QoS-based contention periods following changes made at the iSLA level regarding the environment of this gateway. On the other hand, a self-optimizing function performed by QBAIoT overcomes the problem of unused slots by IoT objects in a QoS CAP. This function is provided by a slot reallocation mechanism covering the entire superframe. The two functions, self-configuring and self-optimizing, make it possible to improve the QBAIoT performance in an IoT environment.

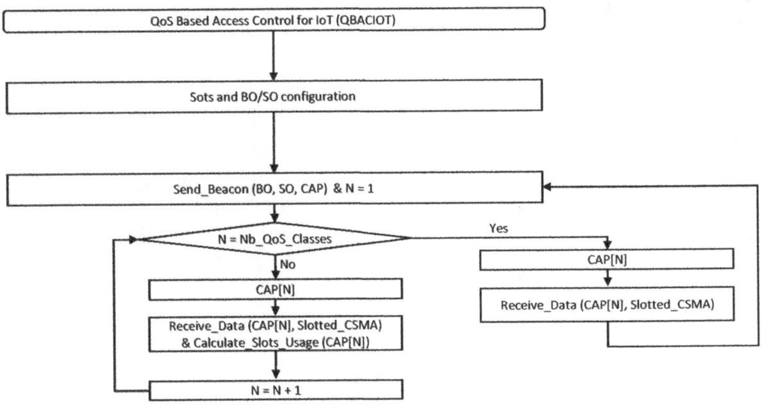

Figure 1.5. *Algorithm for the QBAIoT access method at the gateway level*

1.6.2.2. *QBAIoT objects*

Figure 1.6 depicts the schema for the QBAIoT algorithm used by IoT objects, taking into account the slotted CSMA/CA method.

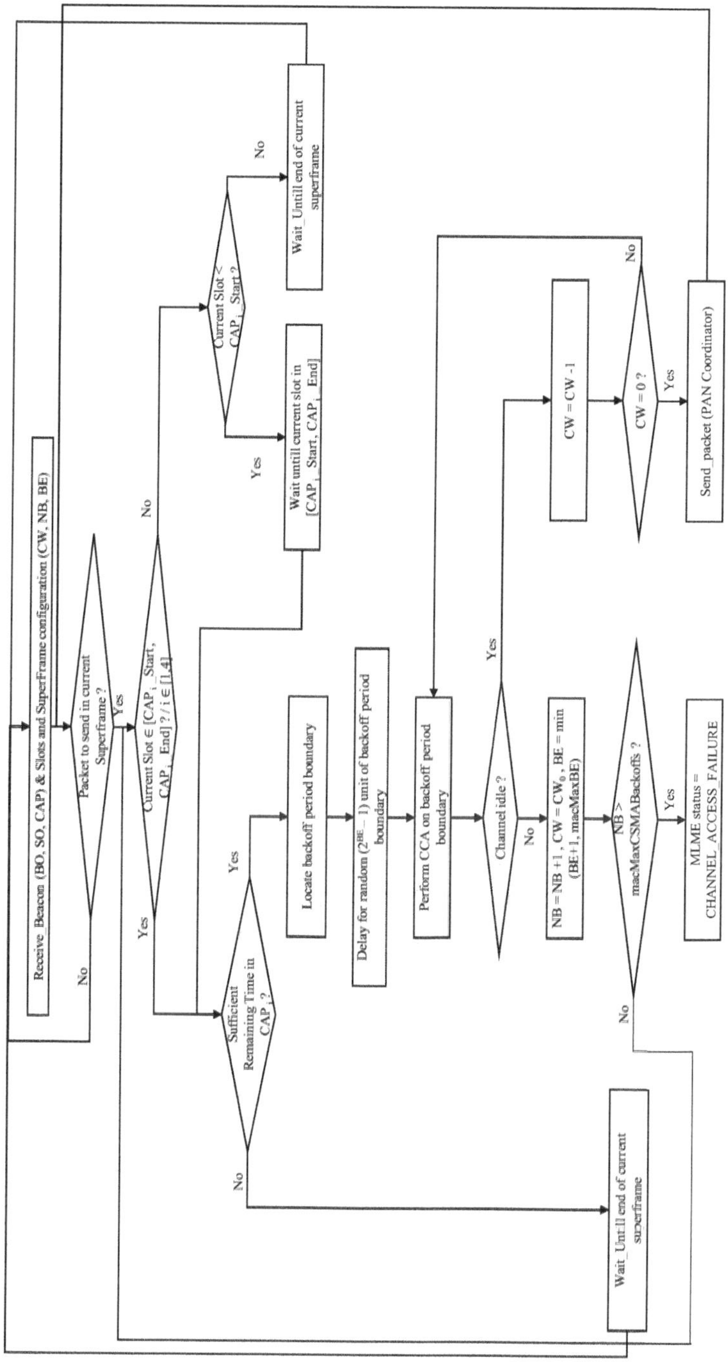

Figure 1.6. *Algorithm for the QBAIoT access method at the IoT object level*

All objects in the environment of the QBAIoT gateway receive the beacon frame. Depending on the QoS class to which they belong, each object determines the QoS CAP during which it can attempt to access the shared channel. When an object generates data, it must test whether it has the right to attempt to send its traffic. If the QoS CAP corresponding to the IoT object in the superframe has not yet started, then it waits for the start of its QoS CAP and then begins competing with other objects in its QoS class to send its data according to the slotted CSMA/CA. On the other hand, if the QoS CAP of the object in the current superframe has elapsed, then it waits for the corresponding QoS CAP in the next superframe. Otherwise, the object has the right to enter into contention with the other objects of its class for access to the channel according to the slotted CSMA/CA mechanism, as its QoS CAP is in progress.

1.6.3. *QBAIoT performance evaluation*

1.6.3.1. *Simulation environment*

As part of the QBAIoT performance evaluation, we use the OMNeT++ network simulator by adapting an IEEE 802.15.4 model available for OMNeT ++ (Kirsche 2018). Adaptation consists of not only deleting the non-contention period (CFP) and the inactive part of the standard superframe but also creating different QoS CAPs within a single superframe. Different simulation scenarios are carried out by varying the number of QoS classes in the considered IoT environment. Table 1.1 describes the configuration and characteristics of the different simulation scenarios in terms of the number of existing QoS classes, the number of objects per class and the distribution of slots by class.

	RTMC objects	RTNMC objects	Streaming objects	NRT objects	Configuration of the slots
Scenario 1	3	0	0	0	16
Scenario 2	3	3	0	0	9/7
Scenario 3	3	3	3	0	7/6/3
Scenario 4	3	3	3	3	6/5/3/2

Table 1.1. *Parameters for the QBAIoT simulation scenarios*

The common value for the BO and SO is set to 2 as at least one real-time QoS class exists in all the scenarios considered. The packet size used in the different simulation scenarios is 50 bytes with a generation interval of 0.25 sec. The duration of each simulation is 100 sec, with a star topology based on a coordinator, namely the QBAIoT gateway.

1.6.3.2. *Results*

The first simulation scenario corresponds to an IoT service based on 3 RTMC objects. This results in an allocation of the 16 slots to the three objects. The second scenario corresponds to IoT services that require three RTMC objects and three RTNMC objects. Consequently, nine slots are allocated to RTMC objects and seven to RTNMC objects. The third scenario corresponds to nine objects (three RTMC, three RTNMC and three Streaming) resulting in a 7/6/3 type configuration of the slots. The last scenario corresponds to the existence of 12 objects (three of each QoS class). This results in a 6/5/3/2 type configuration of the slots.

The results illustrated in Figures 1.7 and 1.8 show that QBAIoT offers better average delays than the IEEE 802.15.4 standard to the QoS classes that are sensitive to this parameter, i.e. to the RTMC and RTNMC classes. It must be noted that the delays for packets with QBAIoT (when a unique QoS class exists in the gateway environments) is slightly greater than the standard, but respects the requirements of RTMC traffic. This difference is due to the test carried out on the slots to verify that they belong to the QoS CAP, which is not necessary for the standard traditional approach without QoS.

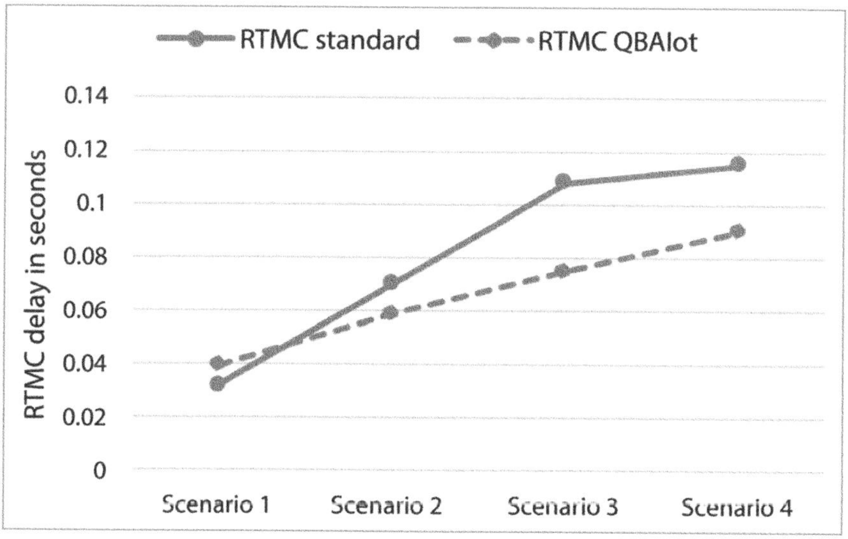

Figure 1.7. *Average real-time mission critical delays for different scenarios*

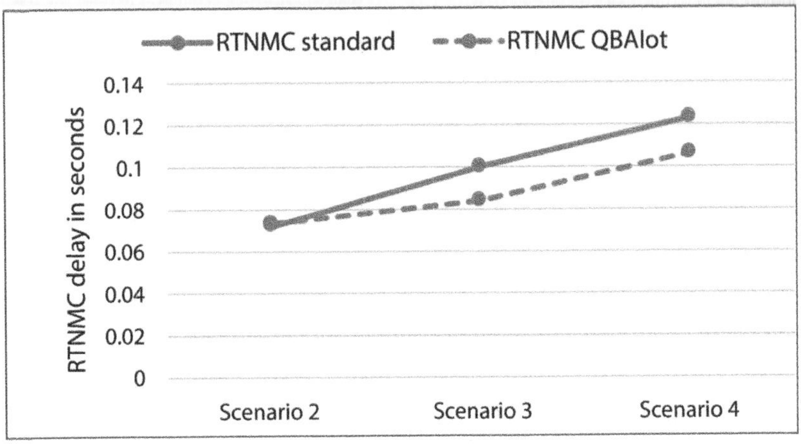

Figure 1.8. *Average real-time non-mission critical for different scenarios*

Figure 1.9 shows that QBAIoT offers better packet delivery ratios for all simulation scenarios and for all QoS classes. Indeed, by limiting the number of objects that can access the channel during a QoS CAP, there are fewer collisions and more packets are successfully transmitted to the destination.

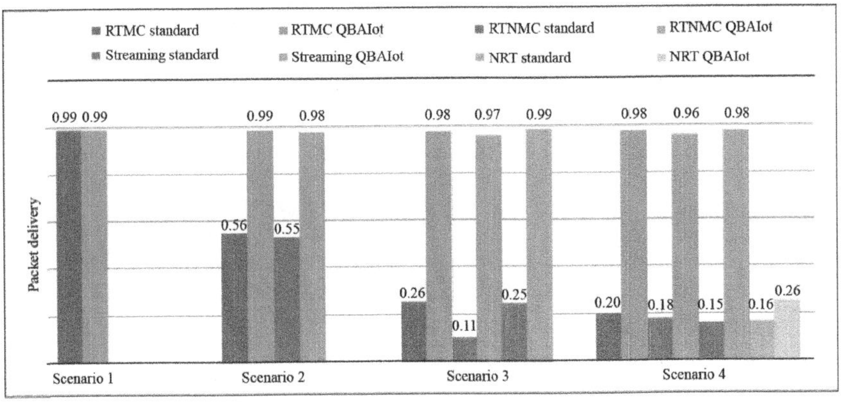

Figure 1.9. *Packet Delivery Ratio for all QoS classe. For a color version of this figure, see www.iste.co.uk/mbarek/service.zip*

1.7. Conclusion

Developments in the domain of the IoT try to optimize the services provided while guaranteeing QoS and security. In this context, research studies and projects and also the various IoT organizations highlight the importance of guaranteeing a service level (QoS and security) between IoT service providers and customers. Different infrastructure, networks and providers interact to offer the IoT service. Thus, research aimed at guaranteeing service levels must take into account the different characteristics of these subenvironments in order to set up mechanisms that are adapted to them in terms of security and QoS.

Security and privacy are essential in the IoT to gain the trust of users and improve the use of services. IoT services impact different areas of daily life and, as a result, the IoT impacts the privacy of users. It is, therefore, essential to comply with European and international regulations concerning privacy protection and this must be integrated into the services offered in an IoT environment. On the other hand, the QoS is another important component of service level that must be considered when offering IoT services. These services can belong to critical fields such as e-health or security. The prioritizing of flows and differentiation of processing are necessary to comply with the requirements and characteristics of these flows corresponding to various fields and services. Several projects have been carried out to set up QoS in the IoT service offering, but there are several challenges that must still be overcome in order to meet the requirements of the IoT environment. In this context, QBAIoT, an approach to guarantee QoS at the device layer of the IoT architecture, makes it possible to differentiate traffic to meet the requirements of each IoT application. This access method based on the IEEE 802.15.4 standard takes into account four categories of service and makes it possible to obtain good performance with a QoS guarantee.

1.8. References

Abdemeziem, M.R. (2016). Data confidentiality in the Internet of Things. PhD Thesis, University of Lorraine, Metz.

AIOTI. (2017). Report on workshop on security and privacy in the hyper connected world. Report, AIOTI.

Alcatel Lucent Enterprise. (2018). The Internet of Things in transportation [Online]. Available at: https://www.al-enterprise.com/-/media/assets/internet/documents/iot-for-transportation-solutionbrief-en.pdf [Accessed 6 March 2020].

Allerin. (2018). Authentication and device identification in IoT security [Online]. Available at: https://www.allerin.com/blog/authentication-and-device-identification-in-iot-security [Accessed 7 November 2018].

ARMOUR. (2016). ARMOUR Experiments and Requirements [Online]. Available at: https://www.armour-project.eu/wp-content/uploads/2016/08/D11-ARMOUR-Experiments-and-Requirements.pdf.

ARMOUR. (2018). Consortium [Online]. Available at: https://www.armour-project.eu/consortium/ [Accessed 2 November 2018].

Babiarz, J., Chan, K., and Baker, F. (2006). Configuration guidelines for DiffServ Service classes [Online]. Memo, IETF, 4594. Available at: https://tools.ietf.org/html/rfc4594.

Balte, A., Kashid, A., and Patil, B. (2015). Security issues in Internet of Things (IoT): A survey. *International Journal of Advanced Research in Computer Science and Software Engineering*, 5(4), 450–455.

Bhaddurgatte, R.C. and Kumar, V. (2015). Review: QoS architecture and implementations in IoT environment. *Research & Reviews: Journal of Engineering and Technology*, S1, 6–12.

Braden, R., Clark, D., and Shenker, S. (1994). Integrated services in the Internet architecture: An overview [Online]. Memo, IETF, 1633. Available at: https://tools. ietf.org/html/rfc1633.

Cantora. (2013). Mirror of AVR-Crypto-Lib. GitHub.

Chowdhury, A., Mukherjee, S., and Souray Banerjee, S. (2018). *Examining of QoS in Cloud Computing Technologies and IoT Services*. Hershey, PA.

CORDIS. (2018). Butler [Online]. Available at: https://cordis.europa.eu/project/rcn/101349/factsheet/en [Accessed 5 November 2018].

Davidson, M. and Schmidt, C. (2014). TAXII overview. Report. The MITRE Corporation, McLean.

Dong, C., Chang, G., Sun, D., Li, J., Jia, J., and Wang, X. (2011). TRM-IoT: A trust management model based on fuzzy reputation for Internet of Things. *Computer Science and Information Systems*, 8(4), 1207–1228.

Duan, R., Chen, X., and Xing, T. (2011). A QoS Architecture for IOT. *International Conference on Internet of Things, and International Conference on Cyber, Physical and Social Computing*. Dalian, China, 717–720.

Ericsson. (2018). Ericsson and partners demonstrate battery life improvements in Massive IoT e-health wearable prototype [Online]. Available at: https://www.ericsson.com/en/news/2018/8/connected-e-health-IoT [Accessed 2 November 2018].

Feng, X., Jie, L., Ruonan, H. *et al.* (2013). Service differentiated and adaptive CSMA/ CA over IEEE 802.15.4 for cyber-physical systems. *The Scientific World Journal*, 2013(947808), 1-12.

Google's Internet of Things Solutions. (2018). Build smart devices with Google [Online]. Available at: https://developers.google.com/iot/ [Accessed 5 July 2018].

Grugen, L., (2018). Updated use cases, requirements and architecture. BigClouT.

GSMA (2016). IoT security guidelines overview document version 1.1. GSMA Association, London.

Gubbi, J., Buyya, R., Marusic, S. *et al.* (2013). Internet of Things (IoT): A vision, architectural elements, and future directions. *Future Generation Computer Systems*, 29, 1645–1660.

Hamze, M. (2015). Autonomie, sécurité et QoS de bout en bout dans un environnement de Cloud Computing. PhD Thesis, University of Burgundy, Dijon.

Hanna, S. (2015). The untrusted IoT: A path to securing billions of insecure devices. Trusted Computing Group.

IEEE Standards Association. (2018). IEEE standard for local and metropolitan area networks – Secure device identity [Online]. IEEE 802.1AR-2018. Available at: https://standards.ieee.org/standard/8021AR-2018.html [Accessed 5 November 2018].

International Electrotechnical Commission (IEC). (2017a). Orchestrating Infrastructure for Sustainable Smart Cities. White book, IEC, Geneva.

International Electrotechnical Commission (IEC). (2017b). IEC Role in the IoT. White book, IEC, Geneva.

International Electrotechnical Commission. (2018). Interconnection of information technology equipment [Online]. IEC, Geneva. Available at: https://www.iec.ch/dyn/www/f?p=103:7:0::::FSP_ORG_ID,FSP_LANG_ID:3399,25 [Accessed 2 November 2018].

ISO/IEC. (2015a). Internet of Things (IoT). Report, ISO, Geneva.

ISO/IEC. (2015b). Information technology, trusted platform module library – Part 1: Architecture. Standard, ISO, Geneva.

ITU-T. (1991). Architecture de sécurité pour l'interconnexion en systèmes ouverts d'applications du CCITT [Online]. Recommendation, ITU. Available at: https://www.itu.int/rec/T-REC-X.800-199103-I/fr.

ITU-T. (2012). Y.2060: Overview of the Internet of Things [Online]. Recommendation, ITU. Available at: https://www.itu.int/rec/T-REC-Y.2060-201206-I.

ITU-T (2014). Y.2066: Next generation networks – Frameworks and functional archi-tecture models. Recommendation, ITU.

Official Journal of the European Union. (2016). Règlement (UE) 2016/679 du Parlement européen et du Conseil du 27 avril 2016 relatif à la protection des personnes physiques à l'égard du traitement des données à caractère personnel et à la libre circulation de ces données, et abrogeant la directive 95/46/CE (règlement général sur la protection des données). Regulation, European Union, Brussels.

Official Journal L 281. (1995). Directive 95/46/EC of the European Parliament and of the Council of 24 October 1995 on the protection of individuals with regard to the processing of personal data and on the free movement of such data. Regulation, European Commission, Brussels.

Khalil, A., Mbarek, N., and Togni, O. (2017). Service level guarantee framework for IoT environments. *First International Conference on Internet of Things and Machine Learning*. IML, Liverpool.

Khalil, A., Mbarek, N., and Togni, O. (2018). IoT service QoS guarantee using QBAIoT wireless access method. *Fourth International Conference on Mobile, Secure, Programmable Networks*. MSPN, Paris.

Khodkari, H., Ghazi-Maghrebi, S., and Asosheh, A. (2017). Assurance of QoS in the Integration of Cloud Services and Internet of Things. *International Symposium on Networks, Computers and Communications*. Marrakech.

Kirsche, M. (2018). IEEE 802.15.4-standalone [Online]. GitHub. Available at: https://github.com/michaelkirsche/IEEE802154INET-Standalone.

Knight, P. (2010). Cryptographic suite for Arduino (SHA, HMAC-SHA) [Online]. GitHub. Available at: https://github.com/Cathedrow/Cryptosuite.

Lee, G.M., Park, J., Kong, N. *et al.* (2012). The Internet of Things: Concept and problem statement [Online]. Memo, IRTF. Available at: https://illumine.wp.imtbs-tsp.eu/files/2012/01/8-draft-lee-iot-problem-statement-00.txt.

Li, S. and Li, D.X. (2017). *IoT Node Authentication. Securing the Internet of Things*. Elsevier, Cambridge.

Lin, S.W., Miller, B., Durand, J. *et al.* (2015). Industrial internet reference architecture version 1.7. Working document, Industrial Internet Consortium, Needham.

LiteOS. (2018). [Online]. Available at: https://www.huawei.com/minisite/ liteos/cn/ [Accessed 5 July 2018].

Menoret, S. (2012). Security requirements for the iCore cognitive management and control framework version 1.0. Document, iCore Project.

Minerva, R., Biru, A., and Rotondi, D. (2015). Towards a definition of the Internet of Things (IoT). *IEEE*, 1.

Mosenia, A. and Jha, N.K. (2017). A comprehensive study of security of Internet-of-Things. *IEEE Transactions on Emerging Topics in Computing*, 5(4), 586–602.

Nagara, K., Aoki, K., Matsubara, Y. *et al.* (2017). Portable DoS test tool for IoT devices. *Workshop on Internet of Things Security and Privacy*. Dallas.

Nef, M.A., Perleps, L., Karagiorgou, S. *et al.* (2012). Enabling QoS in the Internet of Things. *The Fifth International Conference on Communication Theory, Reliability, and Quality of Service*. Chamonix.

Njit. (2018). Smart transportation: How the IoT is changing the transportation system [Online]. Available at: https://graduatedegrees.online.njit.edu/resources/msce/msce-articles/smart-transportation-how-the-iot-is-changing-the-transportation-system/ [Accessed 2 November 2018].

Nokia. (2018). Smart city [Online]. Available at: https://networks.nokia.com/smart-city2018 [Accessed 2 November 2018].

Nordrum, A. (2016). Popular Internet of Things forecast of 50 billion devices by 2020 is outdated [Online]. IEEE Spectrum. Available at: https://spectrum.ieee.org/tech-talk/telecom/internet/popular-internet-of-things-forecast-of-50-billion-devices-by-2020-is-outdated.

Pokric, B., Krimmling, J., Soika-Piotrowska, A. *et al.* (2015). SMARTIE: Node security, policies, and automatic configuration version 1.0. Document, IHP.

Premila, D., Rabara, A., and Jerald, V. (2015). Quality of service architecture for Internet of Things and cloud computing. *International Journal of Computer Applications*, 128(7), 23–28.

Ray, B. (2016). Benefits of Quality of Service (QoS) in LPWAN for IoT [Online]. LinkLabs. Available at: https://www.link-labs.com/blog/quality-of-ser vice-qos-lpwan-iot.

Relic-toolkit. (2018). Relic [Online]. GitHub. Available at: https://github. com/relic-toolkit/relic/commits/master.

Sayarifah, E., Indrajit, A., Sivaramakrishnan, S. *et al.* (2015). An IoT environment for WSN adaptive QoS. *IEEE International Conference on Data Science and Data Intensive Systems*. Sydney.

Serrano, M. (2014). Quality of Service (QoS) for IoT services. Research project, OpenIoT Consortium.

Sicari, S., Rizzardi, A., Grieco, L.A. *et al.* (2015). Security, privacy and trust in Internet of Things: The road ahead. *Computer Networks*, 76, 146–164.

SMARTIE. (2014a). Deliverable D4.1. IoT information Access and Privacy Preservation. Document, IHP.

SMARTIE. (2014b). Deliverable D3.1. Components for secure information gathering and storage. Document, IHP.

Sottile, F., *et al.* (2014). Deliverable D2.5. IoT enabling technologies and future developments. Document, BUTLER Consortium.

TOSSIM. (2018). [Online]. Available at: http://networksimulationtools.com/tossim/ [Accessed 5 November 2018].

Trusted Computing Group. (2018). [Online]. Available at: https://trustedcomputinggroup.org/ [Accessed 5 November 2018].

Windows 10 IoT Core. (2018). [Online]. Available at: https://developer.microsoft.com/en-us/windows/iot/ [Accessed 5 July 2018].

World Health Organization. (2018). WHO and PATH partner to globalize digital health [Online]. Available at: https://www.who.int/ehealth/events/WHO-PATH-partnership/en/ [Accessed 2 November 2018].

Yan, Z., Zhang, P., and Vasilakos, A.V. (2014). A survey on trust management for Internet of Things. *Journal of Network and Computer Applications*, 42, 120–134.

Zarko, V.P., Skocir, P., and Katsaros, K. (2015). Initial report on system requirements and architecture. *SymbIoTe Project*, 1–127.

Ziegler, S. (2015). Deliverable 7.1. Project presentation. Document, F-Interop.

2

Service Level Management in the Cloud

Nader MBAREK
LIB, University of Bourgogne Franche-Comté, Dijon, France

2.1. Introduction

The Internet has developed rapidly over the past 10 years. Small companies with limited resources need to invest on a large scale with an appropriate budget. However, implementing conventional infrastructure is a long and expensive process, which requires a significant initial investment, a lot of time and a large number of qualified personnel to deploy this infrastructure and to manage it.

Additionally, whenever there is a problem with services and applications, these companies need a lot of time and money to resolve them. Consequently, companies that possess their own infrastructure must constantly find the appropriate solutions to these problems. Thus, there arose the question of whether this comprehensive infrastructure needed to be put in place at all, if we could simply rent resources instead. This line of thinking led to the emergence of Cloud Computing.

Although Cloud Computing (or simply, the Cloud) is a relatively recent phenomenon in its current form, we have all probably used the Cloud without being aware of it. Today, Cloud-type environments are omnipresent and growing larger. The main research challenges in this environment are not only service level guarantees, integrating Quality of Service (QoS) and security, but also the management of these services in accordance with a Service Level Agreement (SLA).

Service Level Management in Emerging Environments,
coordinated by Nader MBAREK. © ISTE Ltd 2020.

This chapter will first describe the Cloud environment in terms of Cloud Computing, Cloud Networking, and inter-Cloud. We will then specify the context and motivations concerning the service level guarantee in the Cloud in terms of QoS and security. We will then conclude with a description of a framework for guaranteeing QoS in the Cloud.

2.2. The Cloud environment

2.2.1. *Cloud Computing*

2.2.1.1. *Definition*

The Internet consists of a number of systems that supply services. Cloud Computing, like this, is a set of services and consumable data that is provided to the user (Wygwam 2018). For the majority of users, the Cloud is a new IT model that offers significant savings. It makes it possible to easily respond to development needs and to move away from IT management problems.

Cloud Service Users (CSU) are not usually the owners of the IT infrastructure but are able to access or allocate the Cloud's IT services that are supplied by Cloud Service Providers (CSP).

There are various definitions for the Cloud, with the definition given by the National Institute of Standards and Technology (NIST) (Mell and Grance 2009) touching upon all essential aspects of the Cloud:

> *Cloud Computing* is a model for enabling ubiquitous, convenient, on-demand network access to a shared pool of configurable computing resources (e.g. networks, servers, storage, applications, and services) that can be rapidly provisioned and released with minimal management effort or service provider interaction.

2.2.1.2. *Characteristics*

Cloud Computing has several interesting characteristics that make it an attractive prospect for users (CSU) and service providers (CSP). There are several articles that describe the various characteristics of the Cloud (Gong *et al.* 2010), which include the grouping of shared resources, as well as dynamic and on-demand allocation, ease of access, low initial investment, measured service and advantageous pricing, self-organization and scalability, decreased maintenance costs, virtualized resources and so on.

In a Cloud Computing environment, the goal is to offer users services based on different models. The main service models used in Cloud Computing are as follows:

– Software as a Service (SaaS): This model makes it possible to provide on-demand applications on the Internet. The CSUs neither manage nor control the underlying Cloud infrastructure, including the network, operating systems and storage. A possible exception may be the limited parameters for configuring the application;

– Platform as a Service (PaaS): this model makes it possible to provide developers (CSU) with the resources for execution environments, frameworks for software development and support for operating systems;

– Infrastructure as a Service (IaaS): this model makes it possible to create a pool of virtual resources by sharing physical resources using virtualization technology and can provide these infrastructural resources on-demand, most often in terms of virtual machines (VMs), storage capacity and network resources.

In addition to these, there also exist several models for deployment of the Cloud, which allow providers (CSPs) to offer Cloud services based on the requirements of the user (CSU) in terms of security and QoS. Some examples of this are as follows:

– Public Cloud: in this model, the CSPs offer their resources as services to all kinds of users. This model may be owned, managed and used by a company, an academic or governmental organization or a combination of these;

– Private Cloud: this model of deployment is designed exclusively for a single organization. A private Cloud may be built and managed by the organization or a third-party, or a combination of both;

– Community Cloud: this is a Cloud that shares infrastructure between several organizations within a specific community that have common concerns (goals, security requirements, policies, etc.). The infrastructure is managed by one or more organizations in the community, a third-party or a combination of these entities;

– Hybrid Cloud: this type of deployment is a combination of other Cloud deployment models (public, private and community), which attempts to compensate for the limits of each approach. In a hybrid Cloud, parts of the infrastructure services are provided in private Clouds, while the remaining are hosted on public Clouds. Designing a hybrid Cloud requires determining the optimal distribution between the public and private Cloud.

Several heterogeneous technologies may be used with these different models to create a Cloud Computing environment. Further, several proprietary solutions have been proposed to implement Cloud infrastructure. For this reason, it is essential to bring about standardization in this field, in order to determine the terminology and the techniques to be used, which will allow the emergence of standard, interoperable Cloud infrastructure that can be adopted by different market actors.

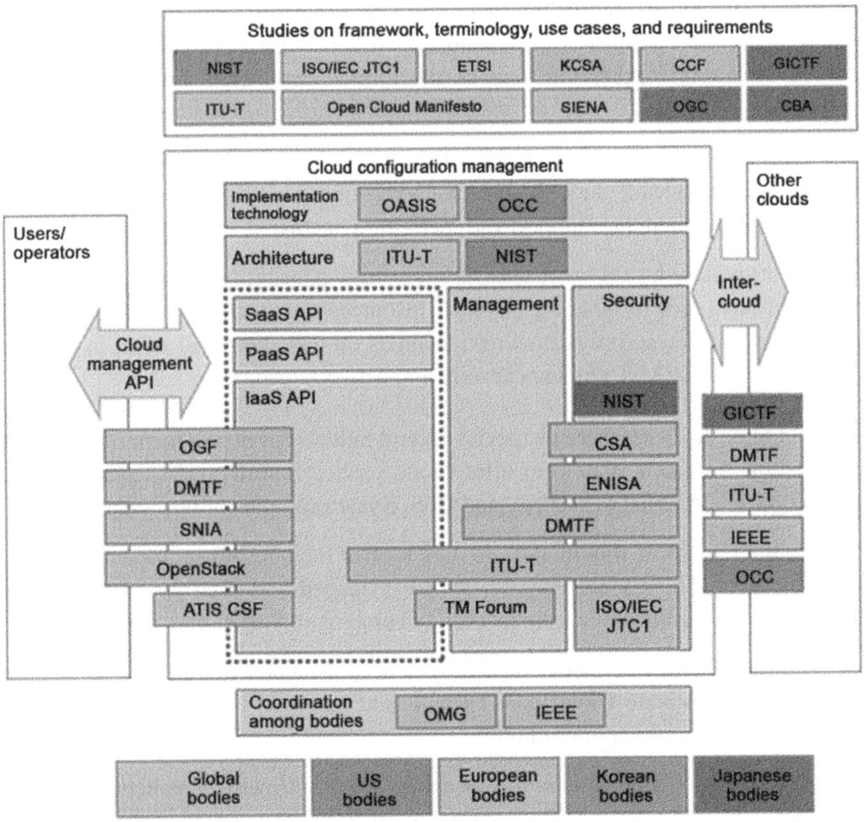

Figure 2.1. *Standardization bodies for the Cloud (Sakai 2011). For a color version of this figure, see www.iste.co.uk/mbarek/service.zip*

Many organizations and informal groups working on Cloud technologies have become specialized in handling standardization problems for the Cloud environment. These standardization bodies help in maintaining norms and best practices to ensure that the different CSPs and equipment are able to coexist in an interoperable manner. Figure 2.1 provides a general classification of the different standardization bodies, based on the roles they play in a Cloud environment.

Several implementation and simulation tools have been designed in order to test Cloud services in a repetitive and controllable environment. These can modify a system before deploying it in the real Cloud and experiment on it with different workloads and performance evaluation scenarios.

Name	Service model	Deployment model	Open source	Goal
OpenNebula (2018)	IaaS	Private Cloud	Yes	Centralized construction and management of IaaS infrastructure
Nimbus (2018)	IaaS	Public Cloud	Yes	Installation on a server cluster and supplying an IaaS service to the CSU
Eucalyptus (2018)	IaaS	Public, Private or Hybrid Cloud	Yes	Installation, deployment and management of server clusters as one single Cloud
OpenStack (2018a)	PaaS, IaaS, NaaS	Public, Private or Hybrid Cloud	Yes	Deployment of Cloud infrastructures and monitoring of resources such as virtual machines, storage or network
CloudStack (2018)	IaaS, NaaS	Public, Private or Hybrid Cloud	Yes	Deployment and management with a web interface of large networks of virtual machines
Snooze (2018)	IaaS	Public or Private Cloud	Yes	Construction of the infrastructure for virtual computing resources and the management of virtual machines

Table 2.1. *Platforms for implementing Cloud Computing*

Table 2.1 describes examples of implementation platforms, whereas Table 2.2 lists simulation softwares for a Cloud Computing environment. These tables summarize different solutions based on the different models of service and deployment that we described in this section.

Name	Service model	Deployment model	Open source	Goal
CloudSim (Calheiros *et al.* 2011)	SaaS, PaaS, IaaS, NaaS	Public, Private or Hybrid Cloud	Yes	Modeling, simulation and transparent experimentation on the Cloud and its services.
RealCloud Sim (2018)	IaaS, NaaS	Public, Private or Hybrid Cloud	Yes	Allocation of VM based on CloudSim and supplying a graphic interface to read BRITE topologies
MR-CloudSim (Jung and Kim 2012)	IaaS	Public, Private or Hybrid Cloud	Yes	Simulation tool based on CloudSim that supports the MapReduce model
GroudSim (Ostermann *et al.* 2010)	SaaS, IaaS	Public, Private or Hybrid Cloud	Yes	Simulation of scientific applications in a Cloud environment.
DCSim (Tighe *et al.* 2012)	IaaS	Public, Private or Hybrid Cloud	Yes	Simulation of a DC (*Data Center*) hosting an IaaS Cloud
GreenCloud (Kliazovich *et al.* 2010)	IaaS, NaaS	Public, Private or Hybrid Cloud	Yes	Simulation of DCs in the Cloud to evaluate their costs in terms of energy. This is an extension of NS2
iCanCloud (Nunez *et al.* 2012)	IaaS	Public, Private or Hybrid Cloud	Yes	Modeling and simulation of Cloud systems. It can provide a global hypervisor and simulate VMs.

Table 2.2. *Simulation software in Cloud Computing*

2.2.2. *Cloud Networking*

2.2.2.1. *Definition*

Cloud Networking offers a new model of Cloud Computing services, namely Network as a Service (NaaS). Because of this new service model, we can use all the resources from the two current types of NaaS (ITU-T 2012). The first network service, that is NaaS (DC), offers network resources from the Data Center (DC) of the Cloud, which connects the various VMs and storage spaces. This kind of network is called an intra-Cloud network. The second network service, NaaS (BoD), offers Bandwidth on Demand (BoD) using network resources that connect the CSU with the CSP infrastructure (basic transport network (WAN/MAN)) or again the network resources that connect the CSP infrastructure to each other (Inter-Cloud network). These network resources may be owned by the same CSP or by different CSPs.

2.2.2.2. *Characteristics*

Cloud Networking has been introduced in a context with multiadministrative domains, where the network and DC must interact via well-defined interfaces to provide a service to the CSU. Cloud Networking makes it possible to specify a service using high-level objectives, while ensuring deployment and rapid management of Cloud services through the DCs and operator networks. This also makes it possible to optimize the management of underlying resources to supply and maintain the required QoS. Cloud Networking extends the virtualization of the network beyond the DC to add two new aspects to Cloud Computing: connecting the user to services offered by the Cloud and interconnecting geographically dispersed services in Cloud infrastructure. Cloud Networking users will be able to ask for their virtual infrastructure and specify the corresponding network properties for accessing these services. There thus exist three types of networks in the Cloud environment (ITU-T 2012):

– intra-Cloud network: this network connects local Cloud infrastructure and includes three subtypes of networks: (a) between VMs on the same server, (b) between servers and storage systems and (c) between VMs on different servers;

– basic transport network (WAN/MAN): this network can be used by clients to access and use Cloud services deployed in the CSP DC;

– inter-Cloud network: this network can be used to interconnect Cloud infrastructures. These Cloud infrastructures may belong to the same Cloud provider or to different providers.

The service offered by Cloud Networking (namely, NaaS) provides network connectivity in the three networks that we just described. It allows users to customize services they receive from the network and allows the provider to decide how resources can be allocated and shared between these users. The requirements of a NaaS-type service include QoS and security. Indeed, it is difficult to guarantee data transfer performance in basic transport networks and inter-Cloud networks. The most practical solution is to use adequate bandwidth to transport data flows in these networks. This bandwidth is generally provided by any entity with the ability to provide network connectivity and a reservation of resources, such as a Network Service Provider (NSP). Thus, the NaaS service offered by these NSPs is BoD. This type of NaaS (BoD) offers the ability to carry out changes on demand (increase bandwidth, decrease bandwidth) and to instantly provide bandwidth over specific links via standardized interfaces. Further, a NaaS (BoD)-type service makes it possible to put in place an end-to-end SLA between the CSU and the CSP. It provides an economical solution for guaranteeing QoS.

Name	Service model	Deployment model	Open source	Goal
OpenStack (2018b)	NaaS	Public Cloud	Yes	Supplying the network resources required for interconnecting the computation and storage OpenStack zones.
Network CloudSim (Garg and Buyya 2011)	SaaS, IaaS, NaaS	Public, Private or Hybrid Cloud	Yes	Modeling the DC Cloud networks and generic applications using CloudSim
CloudNaaS (Benson *et al.* 2011)	SaaS, NaaS	Public Cloud	Yes	A Cloud Networking platform from IBM
Python Open Cloud Networking Interface (Institut Telecom 2018)	NaaS	Private Cloud	Yes	An OCCI extension defined by the Telecom Institute that allows external control of network services
OpenNaaS (2018)	NaaS	Private Cloud	Yes	Supplying a set of flexible tools for deploying NaaS-oriented services

Table 2.3. *Tools for implementing and simulating Cloud Networking*

Several implementation and simulation tools possess features related to Cloud Networking. Examples of such tools are given in Table 2.3.

2.2.3. *Inter-Cloud*

2.2.3.1. *Definition*

If all resources of a CSP have been allocated, then it cannot usually respond to new requests from CSUs. It is also impossible for a CSP to have DCs all over the world. The Cloud must be designed as a multiprovider environment that can enable the migration of a service from one CSP to another and also to localize the best resource, not only in terms of computation capacity or storage capacity, but also in terms of connectivity, bandwidth and forwarding time. In this context, an environment of distributed Clouds (an inter-Cloud environment) is a solution that can respond to the needs of CSUs across various geographical locations (Buyya *et al.* 2010).

The inter-Cloud is, thus, a global interconnection of Clouds – in fact, it could be called the Cloud of Clouds. The inter-Cloud ensures that resources from one Cloud are allocated to others on demand, including resources such as computation capacities, storage and network resources, as well as the transfer of workloads, which is made possible because of the interfunctioning of Cloud systems.

2.2.3.2. *Characteristics*

In the context of an inter-Cloud environment, communication is established between the different Clouds using Cloud Networking, which can provide a NaaS (BoD)-type service. Thus, CSPs may interact in different ways (ITU-T 2012):

– Peer-to-peer inter-Cloud (P2P): this is a direct interconnection between two CSPs using interfaces that have already been established. The two Clouds logically form one single Cloud and one of them may dynamically externalize resources to the other in response to variations in the corresponding demand;

– Inter-Cloud Service Broker (ISB): this model provides an indirect interconnection between two or more CSPs and brokerage services to CSUs or CSPs, through an entity called the ISB, which manages the performances, security and delivery of Cloud services. The ISB provides and executes three categories of services: (a) Service intermediation: The ISB enhances a given service by adding other characteristics to it (access management, security, etc.) and by providing value-added services to CSUs. (b) Aggregation of services: The ISB combines and integrates multiple services into one or more new services. (c) Arbitration of services: This is similar to the aggregation of services, with the exception that the aggregated services are not fixed. In this case, the ISB has the flexibility of choosing services from multiple sources based on a score that allows it to choose the best source;

– Federation-type inter-Cloud: this type of inter-Cloud offers an alliance between multiple CSPs that logically form a single Cloud by integrating their resources. This federation allows CSPs to search for and reserve resources that are available with other CSPs in the alliance, based on different service levels in order to avoid violating SLAs.

Some implementation and simulation tools offer features related to the inter-Cloud and the broker-type inter-Cloud in particular. Examples of these tools are listed in Table 2.4.

Name	Service model	Deployment model	Open source	Goal
IEEE interCloud Testbed (2018)	SaaS, PaaS, IaaS	Public or Private Cloud	Yes	Implementation of a tool for inter-Cloud technologies
Mosaic (2018)	SaaS, PaaS, IaaS	Public or Private Cloud	Yes	A broker Cloud, which allows applications to negotiate the Cloud services requested by their CSUs
Compatible One (2018)	SaaS, PaaS, IaaS	Public, Private or Hybrid Cloud	Yes	A broker Cloud that provides a model for describing, integrating and aggregating different resources provided by heterogeneous CSPs

Table 2.4. *Inter-Cloud implementation and simulation tools*

2.3. Service level and self-management in the Cloud

Service level includes not only QoS but also security, which must be managed autonomously in a Cloud environment. We will first describe the specific features of QoS in the Cloud and we will then study aspects related to security in such an environment. Finally, we will describe how self-management contributes to guaranteeing service levels.

2.3.1. *Quality of Service in a Cloud environment*

2.3.1.1. *Definition and characteristics*

QoS is a management concept that aims to optimize resources and to guarantee applications suitable performance. It is also defined as the capacity to prioritize specific flows and to allocate the necessary resources for responding to the required service levels. Further, QoS denotes a service's capacity to respond through its different characteristics to the various requirements of its CSUs (Kouki *et al.* 2013).

QoS enables a given traffic to be transported in good condition, in terms of several parameters. These QoS parameters may be associated with a NaaS type service (e.g. latency, jitter, packet loss, bandwidth, and availability) or an IaaS-type service (e.g. response time, availability) and also SaaS- and PaaS-type services.

An SLA in a Cloud environment is a bilateral contract between a CSU and CSP or between CSPs. This contract lays out the service conditions and clearly specifies the corresponding service level. The contract may define the QoS required by a CSU and offered by CSPs. When used well, an SLA makes it possible to identify and define client needs, provides the means for understanding and simplifying complex questions, and reduces conflict zones and encourages dialogue in the case of conflict.

The Cloud provides IT resources, software or hardware that can be accessed remotely, as a service. Adopting this model brings with it certain challenges, especially regarding the QoS of the services supplied. However, even today QoS in the Cloud is still not considered in its entirety. While some CSPs have examined QoS criteria, such as the unavailability of a service due to a breakdown, these solutions only look at a single aspect of QoS. As a result, CSUs are invoiced for the quantity of resources used or reserved, but are given only a limited guarantee with respect to the QoS of the resources provided.

Different types of data and traffic for a single CSU have varying levels of importance (critical data, file transfer, real-time communication, etc.). Thus, in order to guarantee the QoS, there must be a mechanism that makes it possible to process these different flows in a differentiated manner, according to the service level corresponding to each. Offering a service that guarantees the values of the QoS parameters in accordance with the preferences of the CSUs is considered to be a significant challenge. The CSP must guarantee delivery of a service with the right QoS to meet the requirements of the clients. There are, therefore, important challenges concerning how to establish an SLA in a Cloud environment and the manner in which the best CSPs are selected for allocating resources based on the CSU's QoS requirements (Yanzhi *et al.* 2013). We must ensure there is continuity in the service level required by the CSU while running their applications.

Further, the resources being deployed in the DCs must have links that allow rapid, flexible and secure connections, not only between the CSU and CSPs, but also between the CSPs themselves. Indeed, many applications require guarantees for bandwidth between user sites and the instances on the servers that execute them. In this context, the SLAs and the provision of the QoS in the Cloud must be effectively standardized in order to enable clear and unambiguous communication between the CSPs and the CSUs.

2.3.1.2. *Research work and standardization*

Many standardization bodies have studied the QoS guarantee in the Cloud environment (as can be seen in Table 2.5).

There are also many research projects related to QoS in Cloud environments, however the QoS in this research is related to SaaS, PaaS and IaaS services, not NaaS services. Consequently, the authors (Patel *et al.* 2009) consider the problem of resource allocation and cost in accordance with an established SLA, but do not consider the inter-Cloud approach or QoS parameters relative to NaaS.

Further, several research projects only introduce an SLA for Cloud Computing, without taking Cloud Networking into account. For example, the Mycloud project (2018) offers a Cloud Service Level Agreement (CSLA). On the other hand, the SAIL (Scalable and Adaptive Internet Solutions) project (SAIL Project 2018) describes a Cloud Networking architecture, and focuses on security, but does not take into account an SLA for QoS. Faniyi *et al.* (2012) present the design for a federated Cloud that allows the coordination of entities in this alliance, in order to respond to the CSU's requirements as described in an SLA. However, they do not consider the case of a broker-type architecture nor any QoS parameters related to NaaS.

Organization	Goal
Distributed Management Task Force (2018)	Working on QoS, SLA, policies, supply and surveillance in the Cloud in the work group "Cloud Management Working Group (CMWG)" created in June 2010
Cloud Standards Customer Council Steering Committee (2018)	Working on QoS and SLA in the document "Practical Guide to Cloud Service Level Agreements Version 1.0"
European Telecommunications Standards Institute (2018)	Working on QoS and SLA in the document "SLA for Cloud service"
Organization for the Advancement of Structured Information Standards (2018)	Working on interactions between the CSP and CSU related to QoS, during design and the supply of services: activities carried out by the technical committee "Symptoms Automation Framework Technical Committee"
SNIA Cloud Data Management Interface (2018)	Working on the standardization of storage and QoS in the Cloud
Open Networking Foundation (2018)	A study of the QoS mechanisms in the next version of the OpenFlow specification

Table 2.5. *Standardization of QoS in the Cloud*

2.3.2. *Security in a Cloud environment*

2.3.2.1. *Definition and characteristics*

Cloud security may be defined as being the set of technical, organizational, legal, and human resources required to protect data, applications and infrastructure associated with the Cloud against a software or hardware vulnerability, which may be exploited by one or more internal or external threats. Cloud security involves concepts such as network and hardware security, as well as control strategies deployed to protect data, applications and infrastructure in the Cloud environment.

The concept of trust in the Cloud may be defined as the CSU's belief that the CSP is capable of providing the required services precisely and infallibly, because of the efficacy of their security mechanisms and their adherence to regulations. Thus, trust is essentially tied to the security processes put in place by the CSP and the Cloud's security model must be based on the hypothesis that the CSU must trust the CSP.

Security offers several important services, such as authentication, access control, confidentiality, integrity, availability and non-repudiation. Consequently, these services become important elements that we must use when designing a secured Cloud system (Zissis and Lekkas 2012).

There may be an SLA around security with respect to the service offered in the Cloud, which defines the CSU's expectations and CSP's obligations to provide Cloud services. Further, a trusted third party may be used to ensure secured interactions and to guarantee trust between the CSU and CSPs or between CSPs, who trust this third party.

In the Cloud, resources are provided as a service that is accessible to clients via the Internet to use on demand as per their needs. These resources are available in the DCs and are accessible from anywhere, such that the Cloud is a single point of access for clients. As a result of this, there were many concerns about trust and security when it came to the transfer, storage, processing, recovery or the destruction of data, as well as computing capacity, processing capacity, storage space and the use of data by a third party in the Cloud environment. Consequently, many organizations were reluctant to externalize their IT systems to the Cloud due to regulatory restrictions and concerns about trust and the guarantee of security.

Security is one of the challenges that must be addressed in order for the Cloud to be adopted by everyone. Externalizing the processing or storage of data in a Cloud Computing environment poses a security risk, as the data may be compromised at different steps in its lifecycle, when it is stored or processed in the Cloud. In addition, security risks may also arise during the transfer of these data from an organization's internal network to the Cloud to be stored or processed, or during data recovery. Thus, the transition toward a large-scale use of Cloud services poses several security challenges, especially with respect to confidentiality. This is chiefly because, in this environment, software and hardware elements that allow a service to be offered belong to multiple trusted domains and it is, therefore, much harder to guarantee security services in a Cloud environment.

2.3.2.2. *Research work and standardization*

Many standardization bodies have studied security in Cloud environments, as seen in Table 2.6.

Organization	Goal
Cloud Standards Customer Council (2018)	Accelerating the adoption of the Cloud by offering norms for security and interoperability linked to transition toward the Cloud.
Global Intercloud Technology Forum (GICTF) (Cloud Standards 2018)	Introducing metrics for SLAs, including security attributes for the inter-Cloud environment.
European Network and Information Security Agency (2018)	Presenting a hierarchical list of organizational, technical, and legal risks with a comparison of the different CSPs based on their security practices.
Cloud Security Alliance (2018)	Proposing several reference documents and guides related to Cloud security.
ITU-T (2018)	Publishing reports and guides on Cloud Computing security.
NIST (Mell and Grance 2009)	Carrying out a project to evaluate security in the Cloud and to study security challenges in the Cloud for the American government.
ISO/IEC JTC1/SC27 (2018)	Studying security and privacy in the Cloud within the WG1/WG4/WG5 working groups
OASIS (2018)	Developing guides for mitigating vulnerabilities and for analyzing risks and threats in the Cloud
SNIA Cloud Data Management Interface (2018)	Developing a standardized interface for security in the Cloud

Table 2.6. *Standardization of security in the Cloud*

There are also several research publications and projects (Neng-Hai *et al.* 2013, Rong *et al.* 2013, Ryan 2013) that present different threats and security challenges that were studied in a Cloud environment. Thus, Christmann *et al.* (2015) describe a method to identify the legal requirements around security for Cloud services. Studnia *et al.* (2012) present vulnerabilities in virtualization, as well as the countermeasures to remedy these vulnerabilities. Ali *et al.* (2015) propose a framework to ensure secured data sharing in an inter-Cloud environment. Fan and Perros (2014) present another framework for trust management in an inter-Cloud environment, but without guaranteeing security services. On the other hand, other research (Sankar 2014) presents architectures that can offer these security services in the Cloud, without looking at the aspects related to trust.

Several research projects have tried to offer security for IaaS Cloud services. For example, Wei *et al.* (2014) describe a framework to guarantee security and privacy for storage and computation resources. Chavan *et al.* (2013) studied the security challenges that were introduced following the evolution toward IaaS Cloud services. Karadsheh *et al.* (2012) studied the risks encountered by implementing the IaaS service model within organizations and described the role of security policies in improving the security of the IaaS service model.

On the other hand, with respect to NaaS Cloud services, Moraes *et al.* (2014) present a framework that makes it possible to isolate CSUs that use the Cloud's network resources. He *et al.* (2014) describe another framework to ensure network security in the Cloud using rule-tree firewalls. Further, for autonomous security management, Ruj *et al.* (2014) present a framework to guarantee autonomous authentication and decentralized access control for data storage in multiple Clouds. Sah *et al.* (2014), however, emphasize integrating and managing authentication, authorization and accounting mechanisms in the Cloud system architecture.

Several research projects have tried to include security in the SLA between the CSU and CSP. Rak *et al.* (2011), for instance, propose constructing an SLA-oriented Cloud application that would enable the management of security functions related to authenticating the user and authorizing the use of services. Chaves *et al.* (2010) give an overview of an SLA that integrates security parameters. Na and Huh (2014) present a methodology for evaluating security based on an SLA and a broker in a Cloud environment. The security mechanisms used when offering a Cloud service must be supervised and managed so as to mitigate security threats. In this context, self-management may be used as an innovative approach to improve the efficacy of the service level guarantee, not only in terms of security but also in terms of QoS, through reducing human intervention in the Cloud environment.

2.3.3. *Self-management of Cloud environments*

2.3.3.1. *Definition and characteristics*

The increase in the complexity, heterogeneity and the dynamic nature of the Cloud environment could lead to fragile, unmanageable and unsecured infrastructure. These characteristics of the Cloud, therefore, today require a more intelligent management strategy, in order to offer services that respond to the demands of the CSU. On the other hand, static management by administrators may lead to various problems with security, scalability, time and cost. We thus need a new management method and there is an urgent need to adopt a new, alternative paradigm, called self-management. In the context of a Cloud environment, the goal of self-management is to develop autonomic systems, whereby applications and resources in the Cloud can be managed in an autonomic manner. This new paradigm makes it possible to better manage the attributes of QoS and security, which often change, due to the dynamic nature of the Cloud, and which must be constantly monitored in order to respect SLAs, minimize violations and reduce penalties.

Self-management seems to be a strategic and holistic approach to designing information systems, applications and complex distributed infrastructures. By adopting this paradigm, these systems can manage themselves in accordance with high-level policies defined by the administrator. Self-management allows a system to adapt its features to changes in its environment in order to improve its performance, its security, its tolerance to failure, its configuration, its maintenance, etc. The concept of self-management, the goal of Autonomic Computing, is inspired by the operation of the human nervous system. The nervous system autonomously determines the actions that it must take in order to maintain a stable state without any external intervention. It thus constantly verifies and optimizes its state and automatically adapts to varying conditions in its environment.

Self-management is characterized by the accomplishment of four management functions in an autonomic manner. These are as follows (IBM 2005):

– self-healing, which denotes the system's capacity to examine, research, diagnose and react in an autonomic manner to malfunctions within the system;

– self-protecting, which is the system's ability to anticipate, identify, detect and protect itself proactively and autonomously against malevolent attacks;

– self-optimizing, which denotes the system's capacity to monitor and autonomously control resources in order to guarantee optimal performance and efficacy with respect to the defined requirements;

– self-configuring, which is the system's ability to autonomously adapt the parameters of the software or hardware resources in order to work correctly or for better performance.

To achieve the goals of self-management, systems are provided with a closed control loop called MAPE-K (Monitor, Analyze, Plan, Execute, Knowledge) that makes it possible to avoid any external intervention. This loop is controlled by an Autonomic Manager (AM) and all these components make up an autonomic system (see Figure 2.2).

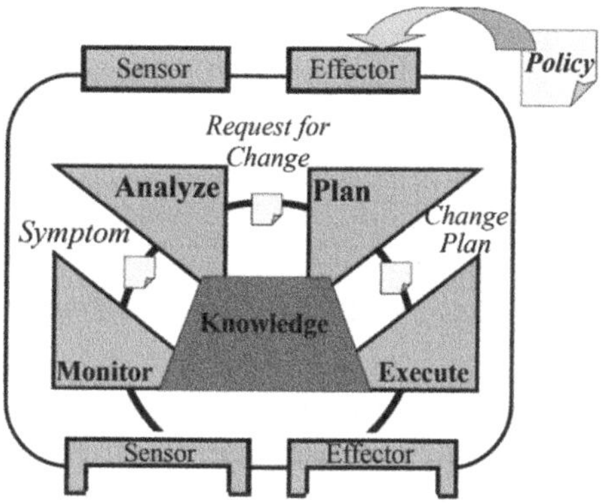

Figure 2.2. *Depiction of an Autonomic Manager (IBM 2005). For a color version of this figure, see www.iste.co.uk/mbarek/service.zip*

An autonomic system is, therefore, composed of an AM, which respects the high-level policies defined by the administrator in order to autonomously manage software or hardware resources using the closed control loop.

The use of Cloud services by different CSUs is steadily increasing, given the advantages that this new technology offers. As part of this increased use, there are a large number of videos transiting between the Cloud and CSUs, and these require an allocation of network resources, adequate bandwidth, several VMs and large storage space. In this context, the management of these resources in a Cloud environment poses a significant challenge in order to satisfy the needs of the CSUs. Managing a Cloud DC involves efficiently handing VMs, storage space and the network in this DC. It also requires managing the network connecting the CSU and the CSP, and the network connecting CSPs to each other. However, a Cloud administrator cannot manage a large number of resources and also ensure that their actions are compatible with the service levels expected at any given moment or ensure that the actions will be compatible with these service levels in the future. This situation, therefore, led to new methods being adopted, which would allow the self-management of Cloud infrastructure, replacing the traditional manual approach.

2.3.3.2. *Research work and standardization*

Several standardization bodies have studied self-management in Cloud environments, as can be seen in Table 2.7.

Name	Goal
Open Networking Foundation (2018)	Adopting the Software-Defined Networking (SDN) technology through open development standards to connect self-management of network resources with QoS demands for Cloud applications.
ITU-T (2018)	Defining a unified platform for the self-management of physical and virtual network devices, as well as managing the Cloud data center.

Table 2.7. *Standardization of security in the Cloud*

Several research studies and projects (Van *et al.* 2009, Infosys 2018) have also presented various mechanisms to implement certain self-management functions in a Cloud environment. Kouki and Ledoux (2012), for example, propose a planning approach that allows a CSP offering a SaaS service to find the optimal configuration for the Cloud application, for a given workload, in accordance with the SLA, while minimizing costs. Patel *et al.* (2009) proposed a mechanism for managing SLAs in a Cloud environment by using Web Service Level Agreement (WSLA), developed for the application and surveillance of an SLA in a Service-Oriented Architecture (SOA). Faniyi *et al.* (2012) describe the design of a middleware for a federated Cloud that could respond to CSU demands with CSP offerings, without violating the SLA. Demchenko *et al.* (2011) propose a security infrastructure for a Cloud that offers IaaS services with a dynamic configuration of the Cloud infrastructure services and SLA management. Ardagna *et al.* (2007) propose an autonomous allocation of resources as well as self-management of SLAs for multiple applications while they are run on a single Cloud. Van *et al.* (2010) define an autonomic resource manager to control the virtualized environment during the supply of resources and the dynamic placement of VMs. The manager aims to optimize a global utility function that integrates both the degree of completion of the SLA as well as the operating costs. Itani *et al.* (2014) describe the design and implementation of ServBGP, an autonomous routing protocol, which aims to find the best route by taking into account performance, security and cost requirements for managing services and collaboration between CSPs in a Cloud environment.

2.4. QoS guarantee in Cloud Networking

2.4.1. *Cloud Networking architectures*

A critical element in providing the expected service level to the CSU is the design and implementation of the Cloud Networking architecture. We have, therefore, proposed two types of architecture for the Cloud environment (Hamze *et al.* 2016): the first is based on the concept of an inter-Cloud broker and the second is based on an inter-Cloud federation. Within these Cloud Networking architectures, we have ensured there is consistency between QoS requirements defined by the CSUs and the SLA proposed by the CSPs.

2.4.1.1. *Inter-Cloud broker-type Cloud Networking architecture*

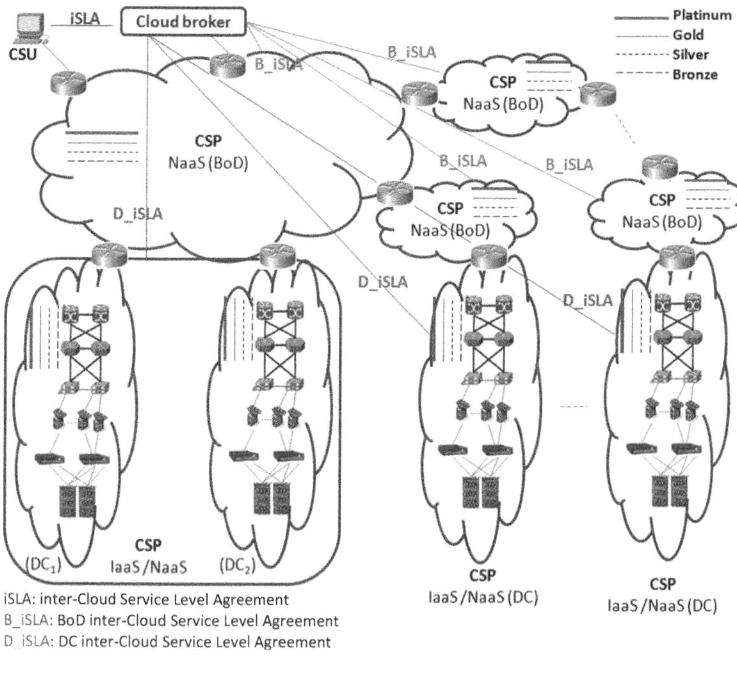

Figure 2.3. *Inter Cloud broker-based Cloud Networking architecture. For a color version of this figure, see www.iste.co.uk/mbarek/service.zip*

In the first architecture proposed (see Figure 2.3), the Cloud broker appears as a flow-through entity between one CSU and many CSPs to help establish a service level that meets the CSU's requirements. It also allows users (such as organizations) to consume a portfolio of integrated Cloud services offered with guarantees that are defined in a single global SLA and with unified invoicing. We have identified two types of CSPs: a CSP (BoD) that provides NaaS (BoD) (the network operator, for example) and a CSP (DC) that provides IaaS and NaaS services. The IaaS service involves VM and storage resources, while the NaaS service has to do with the network resources of the DC. The CSP (DC) may offer resources from one or more DCs. Further, the CSPs may offer different service levels (Platinum, Gold, Silver or Bronze) at different costs and with different guarantees of QoS parameters.

In this context, we have described an SLA that contains the QoS guarantees offered by the CSPs and that are required by the CSUs. Consequently, in our broker-type Cloud Networking architecture, we offer three types of SLAs, created in XML for interoperability and portability between different entities:

1) inter-Cloud Service Level Agreement (iSLA): this is a contract between a CSU and a Cloud broker. It guarantees the QoS for NaaS (BoD and/or DC) and/or IaaS (VM and/or storage). The IaaS QoS parameters, the most important, are Response Time (RT) and Availability ($A_{iaaS.}$), while the NaaS QoS parameters are Latency (L), Jitter (J), Packet Loss Ratio (P), Bandwidth (BW) and Availability (A_{NaaS}). These parameters may also be quantitative or qualitative. The business attributes define the violations, penalties and the unit cost of the VMs, the storage resources and the bandwidth. Thus, the CSU is invoiced depending on their use of the resources and the monitoring interval is used to calculate penalties in the case of a violation of the QoS. Figure 2.4 depicts the most important parameters in this kind of an SLA;

2) BoD inter-Cloud Service Level Agreement (B_iSLA): this is a contract between a Cloud broker and a CSP (BoD). The structure of this SLA resembles that of an iSLA. However, this contract only guarantees QoS for NaaS (BoD) services and only contains quantitative QoS parameters related to NaaS, as well as the unit cost for the bandwidth;

3) DC inter-Cloud Service Level Agreement (D_iSLA): this is a contract between a Cloud broker and a CSP (DC) for NaaS and IaaS services. Its structure resembles that of the iSLA. However, the QoS parameters regarding the IaaS and NaaS services are quantitative.

In the context of our broker-type architecture, establishing an end-to-end SLA requires several interactions between the CSU, Cloud broker and CSP (see Figure 2.5). The CSPs (DC/BoD) first describe the IaaS- and/or NaaS-type services in terms of the resources available with different service levels. The CSPs then periodically send this information to the Cloud broker. However, if there are any significant changes that occur in the CSPs then these modifications are immediately communicated to the Cloud broker.

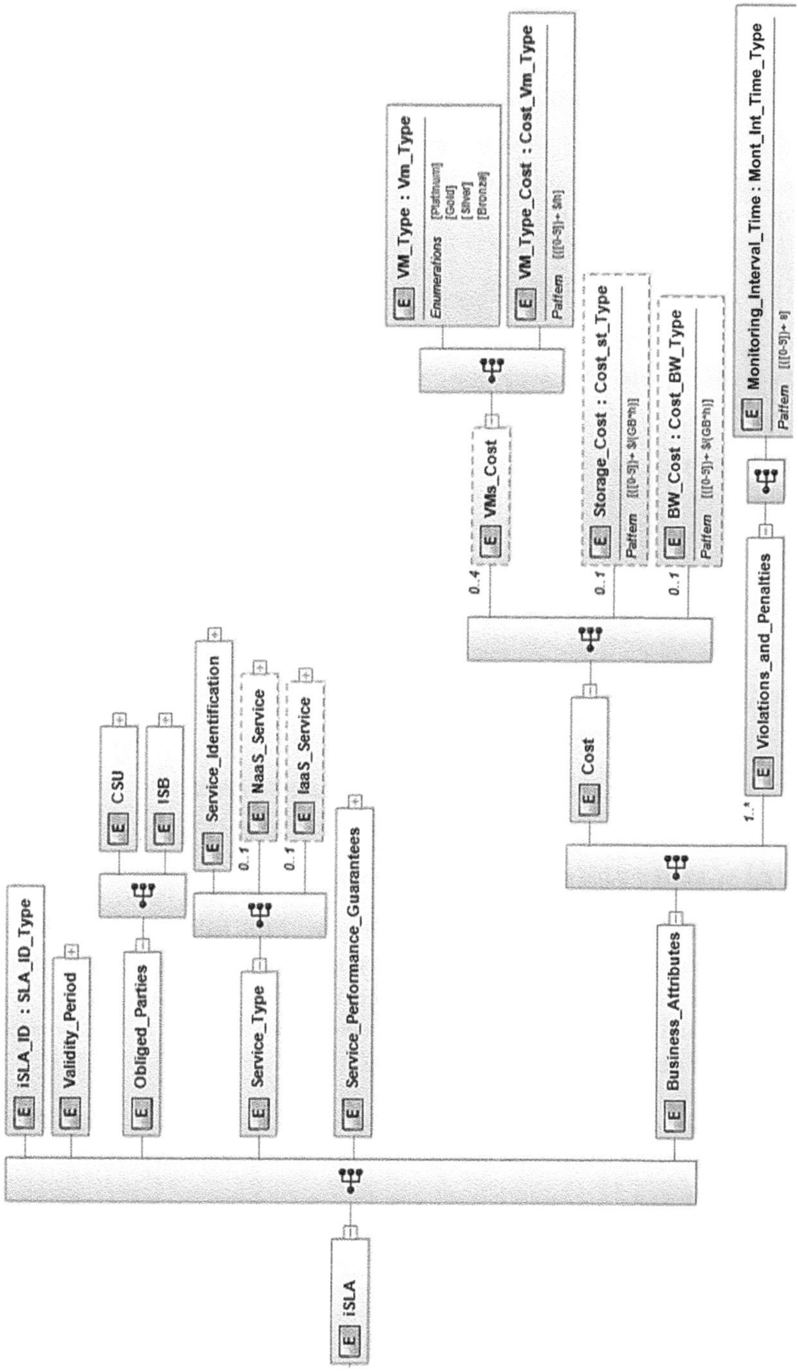

Figure 2.4. *Depiction of the XML diagram for an iSLA*

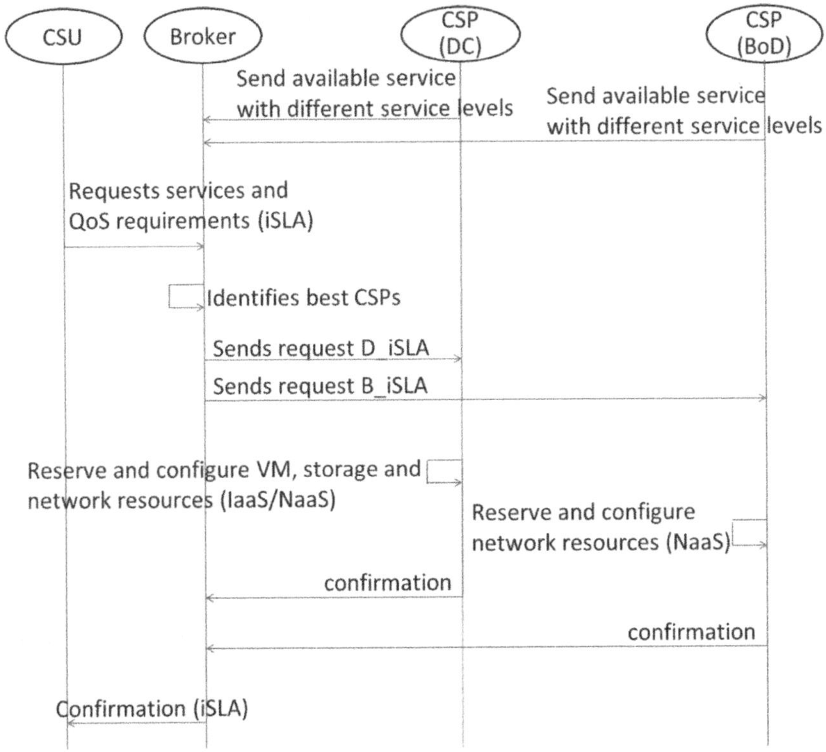

Figure 2.5. *Interactions in a broker-type architecture*

2.4.1.2. *Federated Cloud Networking architecture*

In the second architecture proposed (see Figure 2.6), federation provides an alliance between several CSPs (DC/BoD) who collaborate to help establish a service level that meets the CSU's requirements. The CSPs provide IaaS and/or NaaS services with different service levels. We have considered the CSU to be associated with a CSP (DC) Leader (CSP_L).

We have also proposed the same three kind of SLAs as those for the Cloud broker architecture (see section 2.4.1.1).

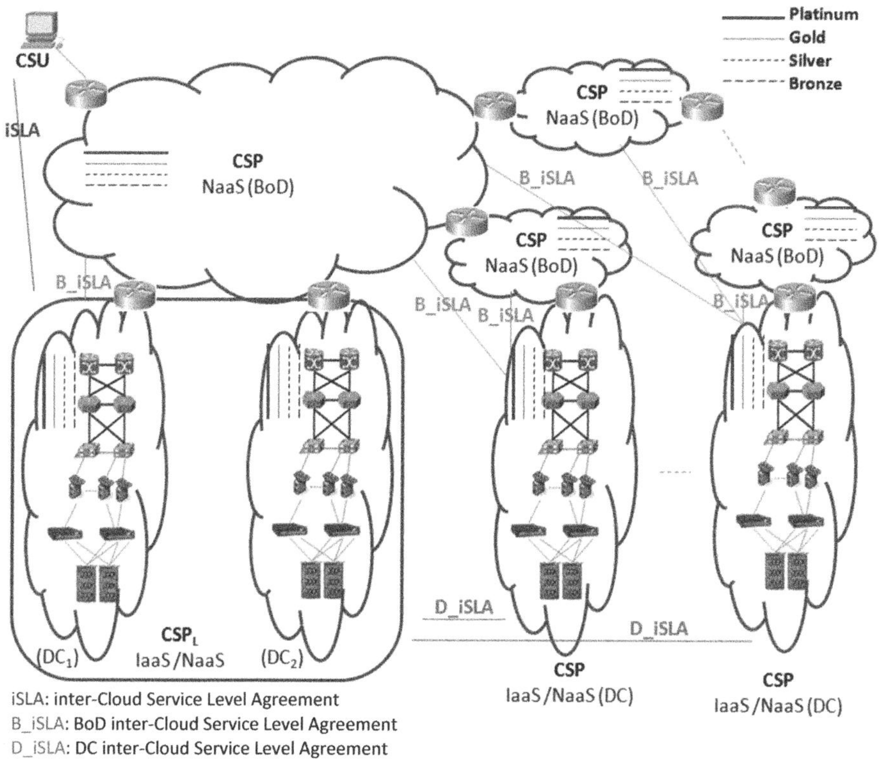

Figure 2.6. *Federation Cloud Networking architecture. For a color version of this figure, see www.iste.co.uk/mbarek/service.zip*

However, the iSLA in this second, federation architecture is a contract between a CSU and the CSP_L, the D_iSLA is a contract between the CSP_L and a CSP (DC), and the B_iSLA is a contract between a CSP (DC) and a CSP (BoD) interconnecting the CSU sites.

Further, in this federation architecture, we have considered two scenarios: in the first, the CSP_L has resources (IaaS with/without NaaS) that respond to the CSU's requirements, while in the second, the CSP_L does not have enough resources and must select the best CSPs that can fulfil the CSU's requirements.

The various interactions between the entities in the federation architecture for these two scenarios are represented in Figure 2.7.

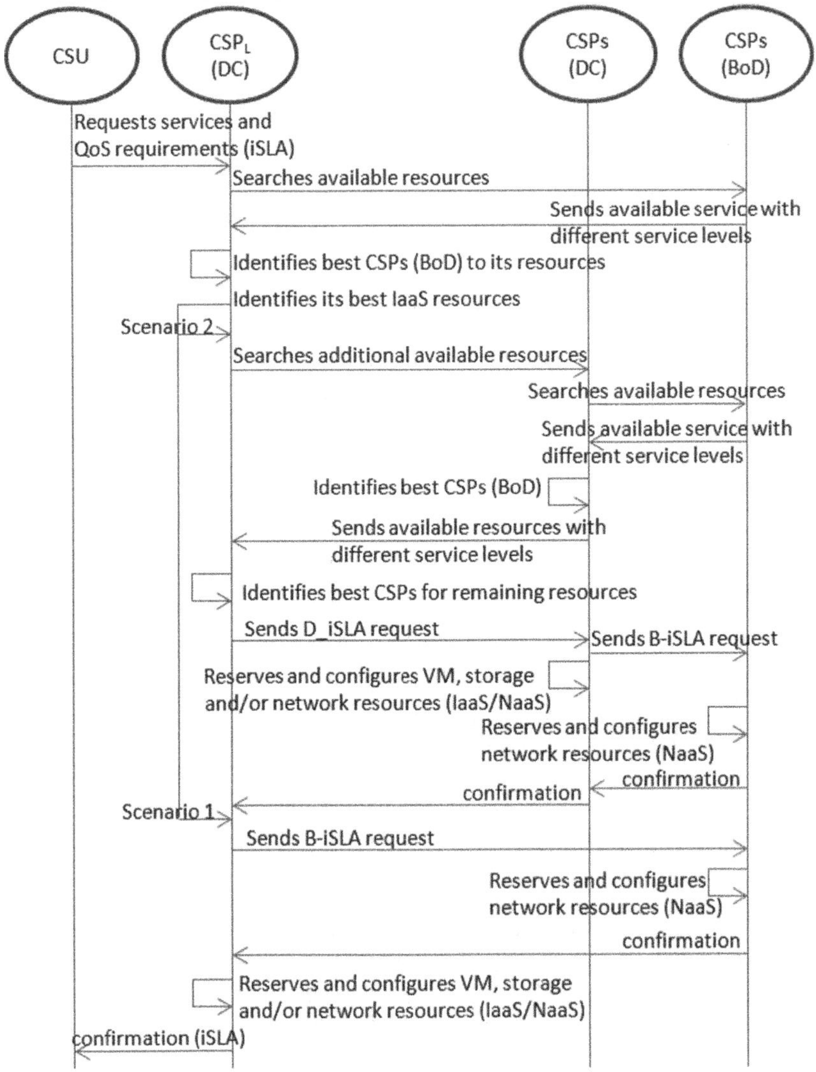

Figure 2.7. *Interactions in a federation-type architecture*

2.4.2. *Performance evaluation*

In order to benefit from the guarantees that would be assured by our Cloud Networking architecture and in order to demonstrate the advantages of architecture of this kind, this section presents two use cases. These will demonstrate the efficiency of videoconferencing applications in the Cloud as well as applications for intensive computing. We have tested the corresponding situation as a proof of concept, allowing us to validate our broker and federation architectures.

2.4.2.1. *Simulation environment*

We evaluated the performance of the two architectures by carrying out a series of simulations corresponding to different scenarios using the CloudSim toolbox (Calheiros *et al.* 2011). We extended CloudSim to support three new entities. The first was a CSU-type entity and the second was a Cloud broker, which replaced CloudSim's DatacenterBroker and allowed the CSU to send their requirements. Finally, the third entity was a CSP (BoD) that interconnected the Cloud broker, the CSP (DC) and the CSU sites, based on a BRITE-type topology (Calheiros *et al.* 2011) to model the bandwidth and the latency of the links. It must be noted that the CSP_L in the federated architecture is simulated as a normal CSP with additional features, such as the selection of the best CSPs.

The simulated model consisted of a Cloud broker or a CSP_L, four CSPs (BoD) and four CSPs (DC), each having 10 hosts. A host has a Quad-core (4 × 1.2 GHz) processor and 16 GB of RAM. All the entities were initiated at the start of the simulation. For a bronze service level, for instance, a CSP (DC) may have the following characteristics in our simulation model: CPU capacity of the VM = 250 MHz, IaaS availability = 98.999%, QoS parameters for NaaS service {Latency = 22 msec, Jitter = 2 msec, Packet Loss Ratio = 10-2, Bandwidth = 1 Mb/sec, NaaS availability = 98.999%}, cost of the bandwidth = 0.005 ($ pet GB) and cost of the processor = 0.11 ($ per hour). In addition to this, a CSP (BoD) may have the following characteristics: NaaS QoS parameters {Latency = 50 msec, Jitter = 5 msec, Packet Loss Ratio = 10-2, Bandwidth = 1 Mb/sec, NaaS availability = 98.999%} and cost of bandwidth = 0.005 ($ per GB). However, the CSPs (DC/BoD) have different values for each service level (platinum, gold, silver and bronze).

2.4.2.2. *Simulation scenarios*

We considered two use cases for the proposed architectures, namely Cloud-based videoconferencing and intensive computing. Multiparty videoconferencing is one of the most demanding multimedia applications in terms of bandwidth, end-to-end delay and jitter. We put in place algorithms to choose the best CSPs in CloudSim. These algorithms determined the best delivery routes for the videoconference flows through the VMs allocated to the videoconference servers, while minimizing the total cost of the application and ensuring there is no violation of any of the corresponding end-to-end QoS constraints.

We evaluated three videoconferencing scenarios. The first corresponded to a static selection of resources with no QoS guarantee. The second and third scenarios corresponded to a selection based on broker and federation architectures, respectively, with a QoS guarantee. There are four CSU sites in each scenario, connected to different CSPs (BoD), with several possible destinations. The CSU chooses the IaaS-type services with NaaS in broker and federation scenarios. In this context, we evaluated the average streaming delay as well as jitter from source to the destination for different scenarios. The corresponding delay is formally defined as the time period required for a packet to travel from an initial point to another, final point and jitter is defined as the variation of this delay.

In the three videoconference scenarios, we simulated four types of videos sent by the CSU. These videos were 1 hour long and the size was 1, 2, 3 and 4 GB, with the required bandwidth being 2.2 Mb/sec, 4.5 Mb/sec, 6.8 Mb/sec and 9.1 Mb/sec, respectively. Additionally, in the broker and federation scenarios, the CSU specifies a maximum network latency of 180 msec, a maximum network jitter of 30 msec and a maximum response time for transcoding of 20 msec. For other QoS parameters related to IaaS and NaaS services, the CSU requires a bronze service level.

2.4.2.3. *Results*

The results illustrated in Figure 2.8 show that a good global delay was obtained because of the CSP selection based on our broker (B) or federation (F) type architecture, compared to static selection (S). The end-to-end delays are well controlled to keep queuing delays to the minimum, while packets traverse through intermediate CSPs on their paths. These CSPs guarantee their part in the overall service level to be ensured for the videoconference flows. Further, jitter is also well controlled and has no impact on the videoconferencing QoS (see Figure 2.9).

As can be seen in Figures 2.10 and 2.11, the costs for the bandwidth and for VMs increases when the size of the video and the corresponding bandwidth increase, due to the IaaS and NaaS-type QoS constraints. In addition, in a static selection, the bandwidth and VM costs are lower than the corresponding costs in the broker or federation scenarios. Finally, the global bandwidth and VM costs in the broker scenario are smaller than in the federation scenario. These results can be explained by the fact that a CSP_L first begins by selecting its own available resource before asking for resources from within other CSPs as needed. However, the resources of the CSP_L are not necessarily the best in terms of cost. We can thus conclude that the broker architecture is the most economical, while also ensuring that the CSU's QoS requirements are met with respect to its real-time application.

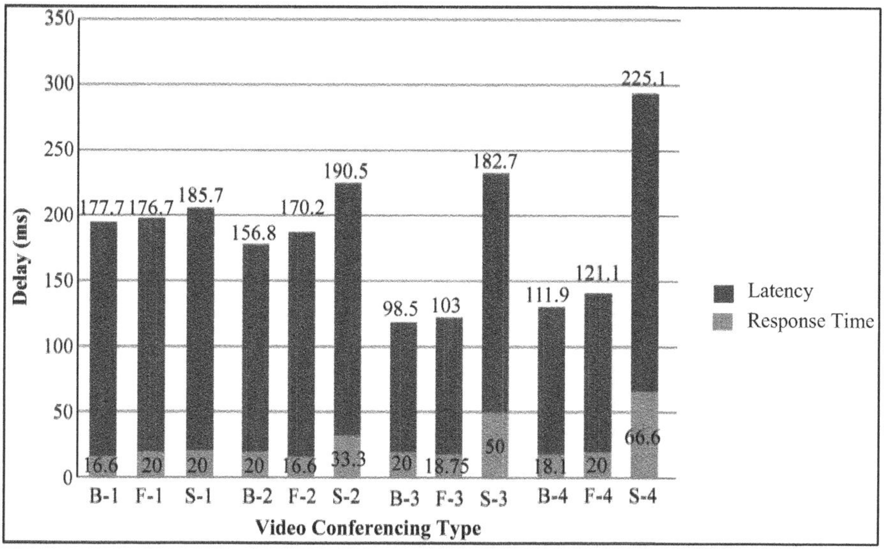

Figure 2.8. *Evaluation of the average global end-to-end delay. For a color version of this figure, see www.iste.co.uk/mbarek/service.zip*

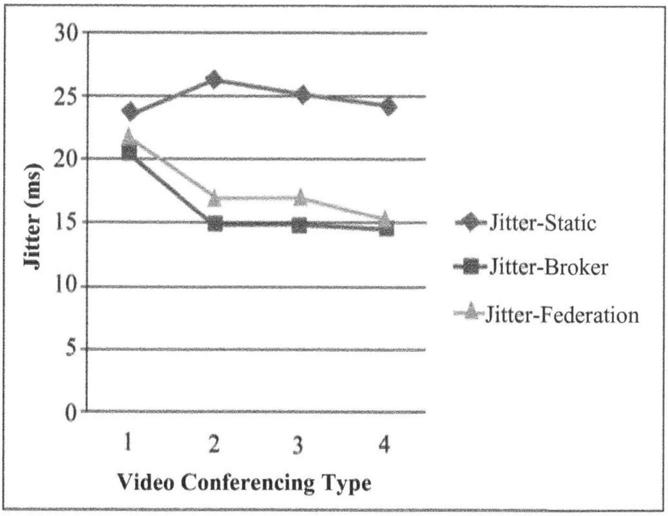

Figure 2.9. *Evaluation of the jitter. For a color version of this figure, see www.iste.co.uk/mbarek/service.zip*

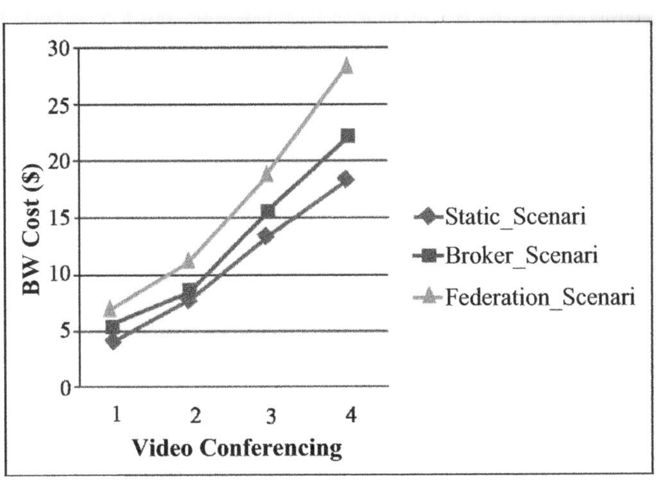

Figure 2.10. *Evaluation of the bandwidth cost. For a color version of this figure, see www.iste.co.uk/mbarek/service.zip*

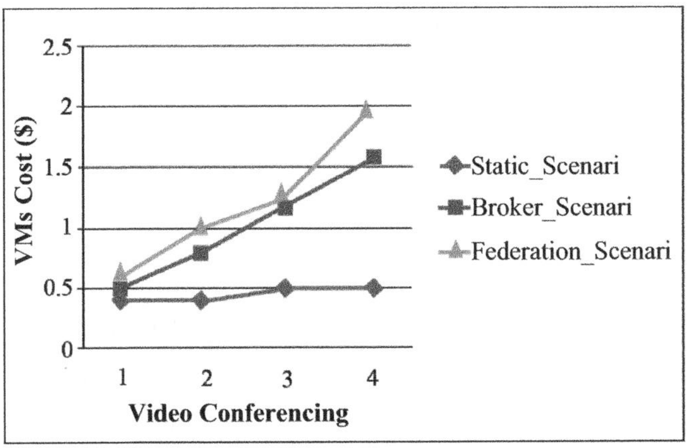

Figure 2.11. *Evaluation of the global VMs cost. For a color version of this figure, see www.iste.co.uk/mbarek/service.zip*

The second use case for our Cloud Networking architecture is an intensive computing application. This may be based on the need to execute specific scientific calculations, for example. In this context, we used the algorithms to select the best CSPs to validate this use case and to determine the best VMs to be allocated to the required jobs, while minimizing the total cost of the application without violating the associated QoS constraints.

We evaluated the response time and latency based on three scenarios. In the first, the CSU chooses an IaaS service without NaaS and a response time lower than 300 msec. In the second scenario, the CSU chooses an IaaS service with NaaS, a response time lower than 300 msec, latency lower than 150 msec, a bandwidth of 1 Mb/sec and the other QoS parameters of the bronze level. In the third scenario, the CSU chooses static selection without a QoS guarantee. In these scenarios, we simulated different job lengths (200, 250, 300 and 350 instructions) run by different VMs. We considered four CSUs, and each sends 100 jobs to four VMs. We assume that one instruction is equal to 1 kB.

As can be seen in Figure 2.12, by providing QoS guarantees for IaaS and NaaS services in the broker architecture (B-I/N) and federated architecture (F-I/N), the response time and latency are well controlled, as compared to only using the QoS guarantee for the IaaS service in both architectures (B-I and F-I) and compared to a static selection (S).

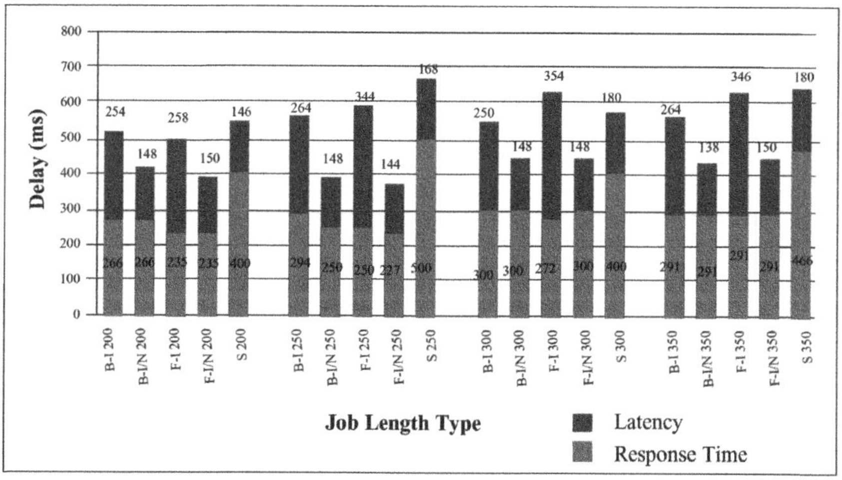

Figure 2.12. *Evaluation of latency and response time. For a color version of this figure, see www.iste.co.uk/mbarek/service.zip*

Further, as can be seen in Figures 2.13 and 2.14, the costs of the bandwidth and for the VMs increase when the job length increases. Additionally, the cost of bandwidth without NaaS (static, B-I and F-I) is lower than the bandwidth cost for B-I/N and F-I/N due to the QoS guarantee for NaaS service. The cost of VMs in static selection is lower than the cost of VMs in the broker and federation scenarios due to the QoS guarantee for IaaS. Finally, the costs of bandwidth and for VMs in the broker scenario are lower than for the federated scenario, for the same reasons given for the videoconferencing use case. Thus, the broker architecture is the most economical of the proposed Cloud Networking architectures, while also guaranteeing a QoS that can satisfy the requirements specified by the CSU.

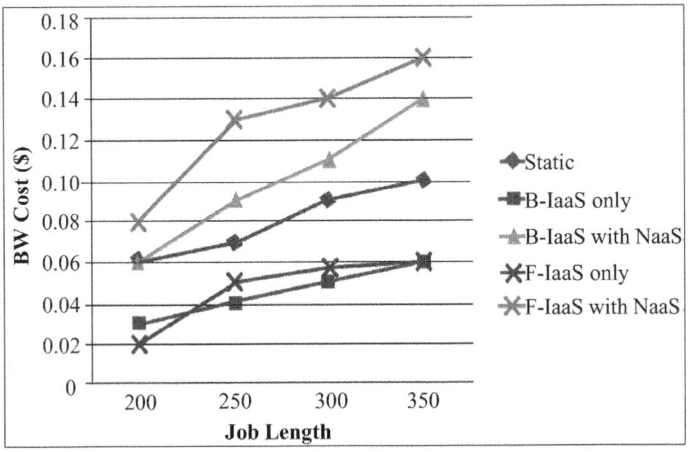

Figure 2.13. *Evaluation of the global bandwidth cost. For a color version of this figure, see www.iste.co.uk/mbarek/service.zip*

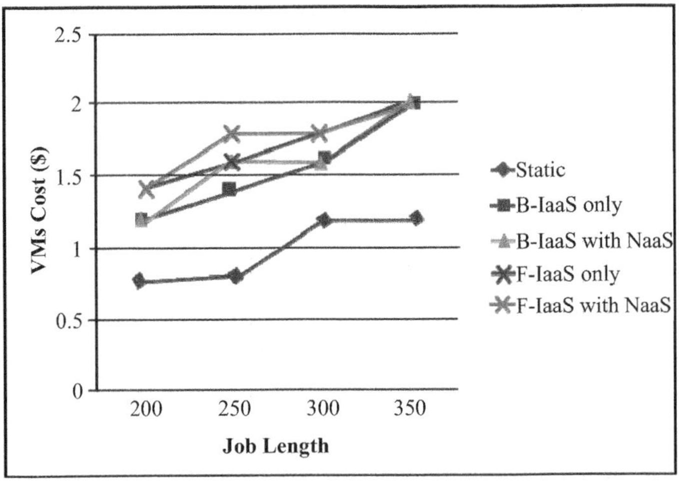

Figure 2.14. *Evaluation of the global VM cost. For a color version of this figure, see www.iste.co.uk/mbarek/service.zip*

2.5. Conclusion

This chapter provides a better understanding of the importance of service level guarantee and self-management in the Cloud environment. We first presented three different types of Cloud environments, namely Cloud Computing, Cloud Networking and the inter-Cloud, emphasizing the characteristics, the most important standardization activities in this type of environment and also the implementation platforms and the simulation tools for the different Cloud types. We then described the importance of the service level guarantee, integrating the QoS and security, as well as self-management in Cloud environments. In this context, we presented standardization activities and research activities that made it possible to overcome certain challenges related to QoS guarantees and security assurance in Cloud environments. We also showed the significance of the concepts of self-management when they are applied in such an environment. In this context, we proposed a framework for the allocation of resources in accordance with an end-to-end SLA between a CSU and several CSP in a Cloud Networking environment (the inter-Cloud broker and federation architectures). The work carried out focused on NaaS and IaaS Cloud services. We also validated the proposed architecture through different simulation scenarios using CloudSim. In the context of these simulations, we also considered real-time applications, such as videoconferencing, in order to provide a QoS guarantee. The results obtained showed that they performed well with respect to the QoS parameters for this type of application. It was especially noted that the broker architecture was the most economical, while also respecting the QoS requirements.

2.6. References

Ali, M., Dhamotharan, R., Khan, E., Khan, S.U., Vasilakos, A.V., Li, K., and Zomaya, A.Y. (2015). SeDaSC: Secure data sharing in clouds. *IEEE Systems Journal*, 11(2), 1–10.

Ardagna, D., Trubian, M., and Zhang, L. (2007). SLA based resource allocation policies in autonomic environments. *Journal of Parallel Distributed Computing*, 67, 259–270.

Benson, T., Akella, A., Shaikh, A., and Sahu, S. (2011). CloudNaaS: A cloud networking platform for enterprise applications. *Second ACM Symposium on Cloud Computing (SOCC'11)*, 1–13.

Buyya, R., Ranjan, R., and Calheiros, R. (2010). Intercloud: Utility-oriented federation of cloud computing environments for scaling of application services. *Algorithms and Architectures for Parallel Processing, 10th International Conference, Proceedings. Part I.* 13–31, 21–23 May, Busan, Korea.

Calheiros, R., Ranjan, R., Beloglazov, A., De Rose, C., and Buyya, R. (2011). CloudSim: A toolkit for modeling and simulation of cloud computing environments and evaluation of resource provisioning algorithms. *Software: Practice and Experience*, 41(1), 23–50.

Chavan, P., Patil, P., Kulkarni, G., Sutar, R., and Belsare, S. (2013). IaaS cloud security. *MIRA 2013: Machine Intelligence Research and Advancement International Conference*, 549–553, 27–29 December, Katra, India.

Chaves, S., Westphall, C., and Lamin, F.R. (2010). SLA perspective in security management for cloud computing. *Sixth International Conference on Networking and Services*. 212–217, 7–13 March, Cancùn, Mexico.

Christmann, C., Falkner, J., Horch, A., and Kett, H. (2015). Identification of IT security and legal requirements regarding cloud services. *Sixth International Conference on Cloud Computing, GRIDs, and Virtualization*. 1–7, 23–28 March, Nice, France.

Cloud Security Alliance (2018). [Online]. Available at: https://cloudsecurityalliance.org/ [Accessed March 2018].

Cloud Standards (2018). Global inter-Cloud technology forum [Online]. Available at: http://cloud-standards.org [Accessed March 2018].

Cloud Standards Customer Council (2018). [Online]. Available at: http://www.cloud-council.org [Accessed March 2018].

Cloud Standards Customer Council Steering Committee (2018). Practical guide to Cloud Service Level Agreements Version 2.0 [Online]. Available at: https://www.omg.org/cloud/deliverables/CSCC-Practical-Guide-to-Cloud-Service-Agreements.pdf [Accessed March 2018].

CloudStack (2018). [Online]. Available at: http://cloudstack.apache.org/ [Accessed March 2018].

CompatibleOne (2018). [Online]. Available at: http://www.compatibleone.org/ [Accessed March 2018].

Demchenko, Y. *et al.* (2011). Security infrastructure for on-demand provisioned cloud infrastructure services. *Third IEEE International Conference on Cloud Computing Technology and Science*. 255–263, 29 November–1 December, Athens, Greece.

Distributed Management Task Force (2018). [Online]. Available at: http://dmtf.org/standards/cloud [Accessed March 2018].

Eucalyptus (2018). [Online]. Available at: https://www.eucalyptus.cloud/ [Accessed March 2018].

European Network and Information Security Agency (2018). [Online]. Available at: https://www.enisa.europa.eu/ [Accessed March 2018].

European Telecommunications Standards Institute (2018). [Online]. Available at: http://www.etsi.org/index.php/technologies-clusters/technologies/grid-and-cloud-computing [Accessed March 2018].

Fan, W. and Perros, H. (2014). A novel trust management framework for multi-Cloud environments based on trust service providers. *Knowledge-Based Systems*, 70, 392–406.

Faniyi, F., Bahsoona, R., and Theodoropoulos, G.A. (2012). Dynamic data-driven simulation approach for preventing service level agreement violations in Cloud federation. *International Conference on Computational Science*. 1167–1760, 4–6 June, Omaha, USA.

Garg, S.K. and Buyya, R. (2011). NetworkCloudSim: Modeling parallel applications in Cloud simulations. *Fourth IEEE International Conference on Utility and Cloud Computing*. 105–113, 5–8 December, Melbourne, Australia.

Gong, C., Liu, J., Zhang, Q., Chen, H., and Gong, Z. (2010). The characteristics of Cloud Computing. *39th International Conference on Parallel Processing Workshops*. 275–279, 13–16 September, San Diego, USA.

Hamze, H., Mbarek, N., and Togni, O. (2016). Broker and federation based cloud networking architecture for IaaS and NaaS QoS guarantee. *13th Annual IEEE Consumer Communications & Networking Conference*. 705–710, 9–12 January, Las Vegas, USA.

He, X., Chomsiri, T., Nanda, P., and Tan, Z. (2014). Improving Cloud network security using the tree-rule firewall. *Future Generation Computing System*, 30, 116–126.

IBM (2005). An architectural blueprint for autonomic computing, 3rd edition. Technical report, IBM, Hawthorne [Online]. Available at: http://www-03.ibm.com/autonomic/pdfs/AC%20Blueprint%20White%20Paper%20V7.pdf.

IEEE InterCloud Testbed (2018). [Online]. Available at: http://www.intercloudtestbed.org/ [Accessed March 2018].

Infosys (2018). FoSII: Foundations of self-governing infrastructures. [Online]. Available at: http://www.infosys.tuwien.ac.at/linksites/FOSII/index.html [Accessed March 2018].

Institut Telecom (2018). PyOCNI [Online]. Available at: http://occi-wg.org/2012/02/20/occi-pyocni/ [Accessed March 2018].

ISO/IEC JTC 1 (2018). [Online]. Available at: http://www.iso.org/iso/jtc1_home.html [Accessed March 2018].

Itani, W., Ghali, C., Bassil, R., Kayssi, A., and Chehab, A. (2014). ServBGP: BGP-inspired autonomic service routing for multi-provider collaborative architectures in the cloud. *Elsevier Future Generation Computer Systems*, 32, 99–117.

ITU-T (2012). Focus group on Cloud Computing. Technical report, ITU, parts 1 to 7.

ITU-T (2018). [Online]. Available at: http://www.itu.int/en/ITU-T/focusgroups/cloud/Pages/default.aspx [Accessed March 2018].

Jung, J. and Kim, H. (2012). MR-CloudSim: Designing and implementing MapReduce computing model on CloudSim. *ICTC 2012: International Conference on ICT Convergence*. 504–509, 15–17 October, Jeju, Korea.

Karadsheh, L. (2012). Applying security policies and service level agreement to IaaS service model to enhance security and transition. *Computers & Security*, 31(3), 315–326.

Kliazovich, D., Bouvry, P., and Khan, S. (2010). GreenCloud: A packet-level simulator of energy-aware cloud computing data centers. *IEEE Global Telecommunications Conference*. 1–5, 6–10 December, Miami, USA.

Kouki, Y. and Ledoux, T. (2012). SLA-driven capacity planning for Cloud applications. *4th IEEE International Conference on Cloud Computing Technology and Science (CloudCom)*. 135–140, 3–6 December, Taipei, Taiwan.

Kouki, Y., Ledoux, T., Serrano, D., Bouchenak, S., Lejeune, J., Sopena, J., Arantes, L., and Sens, P. (2013). SLA et qualité de service pour le Cloud Computing. *Conférence d'informatique en parallélisme, architecture et système*. 15–18 January, Grenoble, France.

Mell, P. and Grance, T. (2009). NIST definition of Cloud Computing v15 [Online]. Available at: https://www.nist.gov/sites/default/files/documents/itl/cloud/cloud-def-v15.pdf [Accessed March 2018].

Moraes, H., Nunes, R., and Guedes, D. (2014). DCPortalsNg: Efficient isolation of tenant networks in virtualized datacenters. *Thirteenth International Conference on Networks*. 230–235, 23–27 February, Nice, France.

Mosaic (2018). [Online]. Available at: http://www.mosaic-cloud.eu [Accessed March 2018].

Mycloud (2018). [Online]. Available at: http://mycloud.inrialpes.fr/ [Accessed March 2018].

Na, S.H. and Huh, E.N. (2014). A broker-based cooperative security-SLA evaluation methodology for personal Cloud Computing. *Security and Communication Network*, 8(7), 1318–1331.

Neng-Hai, Y., Hao, Z., and Xu, J. (2013). Review of Cloud Computing security. *Acta Electron Sinica*, 41(2), 371–381.

Nimbus (2018). [Online]. Available at: http://www.nimbusproject.org/ [Accessed March 2018].

Nunez, A., Vazquez-Poletti, J.L., Caminero, A.C., Castane, G.G. *et al.* (2012). iCanCloud: A flexible and scalable Cloud infrastructure simulator. *Journal of Grid Computing*, 1, 185– 209.

Open Networking Foundation (2018). [Online]. Available at: https://www.opennetworking.org/ [Accessed March 2018].

OpenNaaS (2018). [Online]. Available at: http://www.opennaas.org/overview/ [Accessed March 2018].

OpenNebula (2018). [Online]. Available at: http://opennebula.org/ [Accessed March 2018].

OpenStack (2018a). [Online]. Available at: http://www.openstack.org/ [Accessed March 2018].

OpenStack (2018b). wiki/NaaS [Online]. Available at: https://wiki.openstack.org/wiki/NaaS [Accessed March 2018].

Organisation for the Advancement of Structured Information Standards (2018). [Online]. Available at: https://www.oasis-open.org/committees/ [Accessed March 2018].

Ostermann, S., Plankensteiner, K., Prodan, R., and Fahringer, T. (2010). GroudSim: An event-based simulation framework for computational grids and clouds. In *Euro-Par 2010, Parallel Processing Workshops*, Guarracino, M.R. *et al.* (eds). 305–313, Springer, Berlin.

Patel, P., Ranabahu, A., and Sheth, A. (2009). Service level agreement in Cloud Computing. Report, CORE Scholar, Wright State University.

Rak, M., Liccardo, L., and Aversa, R. (2011). A SLA-based interface for security management in Cloud and GRID Integrations. *7th International Conference on Information Assurance and Security.* 378–383, 5–8 December, Melaka, Malaysia.

RealCloudSim (2018) [Online]. Available at: http://sourceforge.net/projects/realcloudsim/ [Accessed March 2018].

Rong, C., Nguyen, S.T., and Jaatun, M.G. (2013). Beyond lightning: A survey on security challenges in Cloud Computing. *Computer & Electrical Engineering*, 1, 47–54.

Ruj, S., Stojmenovic, M., and Nayak, A. (2014). Decentralized access control with anonymous authentication of data stored in Clouds. *IEEE Transactions Parallel and Distributed Systems*, 25(2), 384–394.

Ryan, M.D. (2013). Cloud computing security: The scientific challenge, and a survey of solutions. *Journal of System Software*, 9, 2263–2268.

Sah, S.K., Shakya, S., and Dhungana, H. (2014). A security management for Cloud based applications and services with diameter-AAA. *IEEE International Conference on Issues and Challenges in Intelligent Computing Techniques*. 6–11, 7–8 February, Ghaziabad, India.

SAIL Project (2018). [Online]. Available at: http://www.sail-project.eu/ [Accessed March 2018].

Sakai, H. (2011). Standardization activities for Cloud Computing. *Global Standardization Activities*, 9(6), 1–6.

Sankar, K., Kannan, S., and Jennifer, P. (2014). On-demand security architecture for Cloud Computing. *Middle-East Journal of Scientific Research*, 20(2), 241–246.

SNIA Cloud Data Management Interface (2018). [Online]. Available at: http://www.snia.org/cdmi [Accessed March 2018].

Snooze (2018). [Online]. Available at: http://snooze.inria.fr/ [Accessed March 2018].

Studnia I., Alata, E., Deswarte, Y., Kaâniche, M., and Nicomette, V. (2012). Survey of security problems in Cloud Computing virtual machines. *Computer and Electronics Security Applications*, 61–74, 20–22 November, Rennes, France.

Tighe, M., Keller, G., Bauer, M., and Lutfiyya, H. (2012). DCSim: A data center simulation tool for evaluating dynamic virtualized resource management. *The 6th International DMTF Academic Alliance Workshop on Systems and Virtualization Management: Standard and the Cloud*. 385–392, 26 October, Las Vegas, USA.

Van, H., Tran, F., and Menaud, J.M. (2009). SLA-aware virtual resource management for Cloud infrastructures. *The 9th IEEE International Conference on Computer and Information Technology*. 357–362, 11–14 October, Xiamen, China.

Van, H., Tran, F., and Menaud, J.M. (2010). Performance and power management for Cloud infrastructures. *The 3rd IEEE International Conference on Cloud Computing*. 329–336, 5–10 July, Miami, USA.

Wei, L., Zhu, H., Cao, Z. *et al.* (2014). Security and privacy for storage and computation in Cloud Computing. *Information Sciences*, 258, 371–386.

Wygwam (2018). *Le Cloud Computing: réelle révolution ou simple évolution?* [Online]. Available at: https://www.yumpu.com/fr/document/view/6220300/le-cloud-computing-reelle-evolution-ou-simple-wygwam [Accessed March 2018].

Yanzhi, W., Shuang, C., and Pedram, M. (2013). Service level agreement-based joint application environment assignment and resource allocation in cloud computing systems. *IEEE Green Technologies Conference*. 167–174, 4–5 April, Denver, USA.

Zissis, D. and Lekkas, D. (2012). Addressing Cloud Computing security issues. *Future Generation Computer Systems*, 28(3), 583–592.

3

Managing Energy Demand as a Service in a Smart Grid Environment

Samira CHOUIKHI, Leila MERGHEM-BOULAHIA and Moez ESSEGHIR

ICD, University of Technology of Troyes, France

3.1. Introduction

The ever-increasing demand for electrical energy is a major challenge for authorities around the world. However, current electricity networks, set up over the last decade, are rapidly growing old. This infrastructure, inherited from bygone years, is becoming increasingly congested and will be unable to fulfill the future energy needs of a digital economy. Further, in order to reduce the impact of climate change on the environment, there is a need to integrate clean and renewable energy. Thus, these electrical facilities must become smarter, more reliable and more robust in order to take into account these new changes, without losing stability and efficiency; hence the emergence of the Smart Grid environment (Xi 2012).

Moving toward a Smart Grid requires the development of new techniques and strategies to obtain a network that makes optimal use of the features and services offered by the Smart Grid.

Several of the proposed algorithms, therefore, target energy management services, particularly demand management, in smart electrical networks. Both online

Service Level Management in Emerging Environments,
coordinated by Nader MBAREK. © ISTE Ltd 2020.

algorithms (responding to events in real time) and offline algorithms (assuming that all information is available, such as predictions, for example) have been proposed.

These algorithms use optimization techniques such as linear, quadratic, dynamic and stochastic programming, game theory, evolutionary algorithms, etc. to reduce spikes in demand, to balance supply and demand, and to minimize waste and energy prices. Distributed Artificial Intelligence techniques are often used for online algorithms, especially multiagent systems (MAS).

In this book, we shall study Demand Management as a service offered to consumers in a Smart Grid environment. We will first introduce this environment and its essential elements. We will then present the fundamental concepts of demand management and finally, we will focus on demand-side management, specifically with respect to energy consumption scheduling.

3.2. The Smart Grid environment

Smart Grids are electrical networks that integrate information and communication technologies for optimal management of resources and to enhance the services provided. These technologies enable coordination between all actors in the processes of generating, distributing, transmitting and consuming electricity (suppliers, consumers, equipment, etc.). From a technical point of view, a Smart Grid may be divided into three major systems: a smart infrastructure system, a smart management system and a smart protection system.

– *Smart infrastructure system*: this is the basic infrastructure for energy, information and communication in the Smart Grid. The infrastructure supports a bidirectional flow of electricity and information, i.e. electricity may be fed into the network by the consumers. This infrastructure may, in turn, be divided into three subsystems: an *energy system* (generation, distribution and consumption of energy), an *information system* (information collection, surveillance, monitoring and management) and a *communication system* (transmission of information between systems, devices and applications in a Smart Grid context).

– *Smart management system*: this system provides services and advanced features for monitoring and management. As new applications are developed and new management services are supported by smart infrastructure, the network continues to become smarter. The management system may set many goals, generally related to improving energy efficiency, balancing supply and demand, controlling emissions, reducing operating costs and maximizing utility.

– *Smart protection system*: this system provides advanced network reliability analysis, fault protection and security and privacy services. The system makes it possible to develop mechanisms to respond to cybersecurity threats and protect privacy.

3.2.1. *Smart microgrids*

Replacing a traditional electrical network with a Smart Grid (usually by making use of existing facilities and resources from the older network) poses several challenges. Smart microgrids (see Figure 3.1) are part of the proposed solution to overcome problems arising from legacy infrastructure. Their ability to integrate local generators – as a resource for distributed production of energy – and storage systems, offers a number of advantages compared to existing systems.

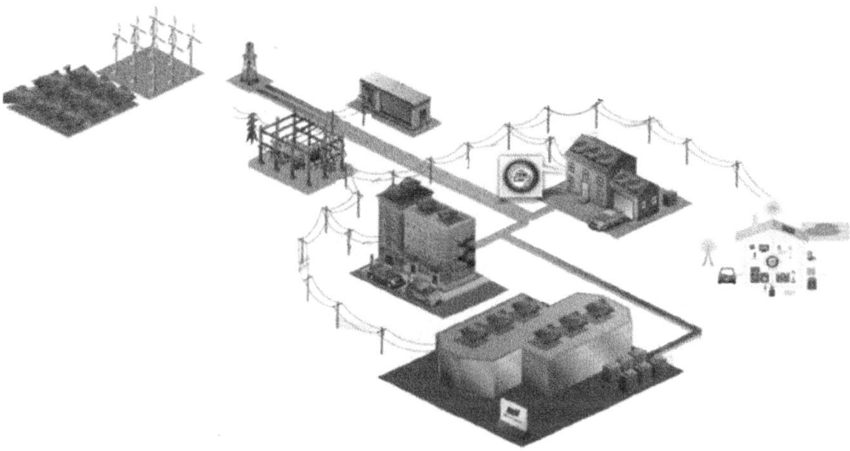

Figure 3.1. *Architecture of a microgrid. For a color version of this figure, see www.iste.co.uk/mbarek/service.zip*

During normal operation, a microgrid is connected to a conventional main grid. Its consumers can generate low-voltage electricity by using distributed resources, such as solar panels, wind turbines and fuel cells. However, the micogrid may be disconnected from the main grid and continue to supply energy to consumers through its distributed power generation resources. Thus, multiple distributed generators, and the ability to isolate the microgrid from any network disruptions, make it possible to have an uninterrupted supply of electricity.

It must be noted that even if users do not get energy from outside in the isolated mode, they can still exchange information with the main grid. It is also proven that the use of microgrids facilitates the implementation of Smart Grid services and features.

3.2.2. *Information and communication infrastructure*

The development of Smart Grids relies not only on advances in technologies and energy equipment, but also on the improvement of automated monitoring, analysis and sophisticated control. These services are distributed and localized. In general, the challenges of distributed automation must be addressed through an information technology perspective, like the interoperability of data and the integration of devices, systems and existing and future applications. In this way, a smart information subsystem is implemented to support the generation, modeling, integration, analysis and optimization of data in a Smart Grid environment. There are many technologies that make up this system, namely smart meters, Phasor Measurement Units (PMUs) and sensors. Given their overall low cost, Wireless Sensor Networks (WSNs) (Akyildiz *et al.* 2002) can offer a feasible and profitable platform for communication and data collection, for monitoring and diagnosis of distributed systems.

However, in order to use this kind of network, the Smart Grid must take several factors into consideration, such as quality of service, limited sensor resources, a hostile environment, security of transmission, etc. To overcome these challenges, the WSNs can make use of techniques and technologies such as the ZigBee protocol, multichannel communication and self-repair features.

In addition to information systems, communication infrastructure is also integrated into the Smart Grid in order to allow the sharing of data which is essential to the functioning of the Smart Grid. Home Area Networks (HANs), in particular, are used to interconnect objects in the consumer's home. A wide range of technology may be used to implement HANs, which may be wired or wireless. Ethernet remains a very popular solution given its high speed, going up to hundreds of gigabits per second. However, this technology requires investment to deploy wired infrastructure. A more promising alternative is the use of Power Line Communication (PLC), which uses cables already installed in the electrical network, instead of installing new cables dedicated to communication. Wireless technologies, such as Wi-Fi (Hiertz *et al.* 2010), are also used. In general, wireless solutions offer communication over short distances, with relatively low speeds, due to attenuation of signals and interference. Other wireless technologies represent a better alternative given that they can create communication infrastructure that is low-cost, flexible, reliable and auto-repair, in case of problems. In addition to HANs, metropolitan Wide Area Networks (WANs) are required in order to connect EMUs to other Smart Grid domains, such as the utility company or the central server, in the case of DSM solutions, designed for cooperative consumer communities. Cellular networks are considered among the most promising technologies given that they can support a large number of simultaneous connections with very wide coverage. In addition, fourth-generation technology such as WiMAX (Andrews *et al.* 2007) and Long-Term Evolution (LTE) (Sesia *et al.* 2011) can guarantee reliability and quality of service during data transmission.

3.3. Demand management: fundamental concepts

Managing the demand for energy is considered one of the most important services offered by a Smart Grid environment. Demand management services are generally classified into two categories: Demand Response (DR) and demand-side management (DSM) services. DR services offer reactive solutions that encourage consumers to dynamically change their demand in the short-term as a result of signals (rate changes, congestion, emergency conditions, etc.) from the system. These techniques are typically used to reduce demand around peak hours or to avoid system failures. On the other hand, DSM services are generally based on proactive services and the goal of these services is to improve the energy efficiency of consumers. In literature, the terms DR and DSM are often used interchangeably or considered synonymous, however they represent methods that are completely different, but which can be used at the same time. In addition to these two types of services, a third type is offered for the Smart Grid environment: predicting energy demands. This service is largely used by the DSM and DR services.

3.3.1. *Predicting loads*

By predicting energy consumption, we can balance demand and the supply of energy, thereby reducing the percentage of unsatisfied consumers. This service makes it possible to guarantee a balance between the demand and supply of energy in order to reduce dissatisfaction. The prediction may be short-term, medium-term or long-term over any period from 1 hour to several years. Predictive approaches generally use artificial intelligence techniques and statistical methods such as exponential smoothing, neuronal networks, expert systems, regression, temporal series and fuzzy logic.

Many algorithms have been proposed to accurately predict demand. Ghasmi *et al.* (2016) modeled the correlation between predicting the price of electricity and predicting demand, through a multi-input multi-output system that uses the ARIMA (Auto-Regressive Integrated Moving Average) methods and NLSSVM (Nonlinear Least Square Support Vector Machine). All the NLSSVM parameters are converted into an optimization problem that is resolved by an artificial bee colony algorithm. Simulation shows that precise predictions are obtained using a hybrid algorithm. A fuzzy, hybrid evolutionary model with parametric optimization was used to predict demand in Coelho *et al.* (2016). Optimization is carried out using a bio-inspired optimizer. Real data are used to validate the methodology and the results show that this solution is more suitable for short-term prediction in microgrids and larger networks.

Results obtained using predictive algorithms are largely used by DSM approaches that manage demand in order to reduce consumption costs, balance supply and demand, smooth the consumption profile, etc.

3.3.2. *DR – demand response*

The implementation of the DR service makes it possible to overcome the problems related to consumption peaks and emergency cases. The proposed mechanisms reward consumers who reduce consumption following a request, or consumers hand over control of some electrical equipment to system administrators. Examples of this incentive system include:

– direct load control: in this solution, the network manager can intervene directly to control the user's consumption. Upon observing a problem in the network, the manager sends a signal to cancel the requests of certain applications, in order to stabilize the system. In Zhu *et al.* (2015), the authors have used this technique as part of their solution for consumption scheduling in a microgrid that has integrated renewable resources and storage units;

– interruptible/curtailable rates: in exchange for lower rates, the user must either reduce their energy demands in the short-term or accept a temporary interruption in the energy supply. This solution is usually used in peak hours when there is heavy energy demand. Aalami *et al.* (2010) introduced an economic model in which energy consumption depends on the price of electricity, elasticity of the demand and the penalty. In order to satisfy the consumers and smooth the demand profile, the network manager may use this model to simulate user behaviors for different prices, elasticities and penalties;

– Emergency Demand Response Programs (ERDP): when there is a risk of the system becoming destabilized, the network manager proposes offering consumers incentives to reduce their demand. Sahebi *et al.* (2012) compared this technique with the interruptible/curtailable rates technique on the same system and demonstrated that ERDP offers better performance in terms of the total cost;

– capacity market programs: clients commit to reducing their demand during peak hours. If they fail to adhere to this when the manager requests them to reduce their consumption, they will be penalized. In exchange for this commitment, these consumers benefit from special rates, even if no reduction in consumption is requested. This technique was used by Aalami *et al.* (2010) to develop the penalty in the model they proposed;

– ancillary services market programs: consumers who commit to reduce their demand, if necessary, are paid by network opearators. This makes it possible to support and maintain the smooth working of the electrical network. Kirby and Hirst (1997) and Trehan (2001) have described services of this type in detail;

– demand bidding/buyback programs: in this technique, it is the consumer who initiates a bidding process with operators for reducing demand. The offer consists of the quantity of energy that the user is willing to reduce and the price they are asking for. This solution is generally advantageous to energy-intensive consumers who can change and adapt their demand. In Saebi *et al.* (2010), the authors studied this technique from the point of view of the system manager and proposed a model that allowed energy demands to be redistributed better.

A new solution is proposed in Holyhead *et al.* (2015) that makes it possible to select an optimal subset of consumers who will be targeted with a request to reduce energy consumption. These consumers are likely to use devices that can be scheduled around peak times and are likely to react positively to the energy supplier's request.

3.4. Demand-side management

In this section, we will examine the basic architecture and components of DSM platforms. We will also study the different approaches (see Table 3.1) and classify them based on different criteria.

DSM systems are designed to identify optimal scheduling for a single residential consumer or for a community of consumers. This scheduling may have several goals: minimizing electricity bills, minimizing inconvenience to the consumer, maximizing the consumption of locally produced energy, maximizing the use of renewable energy, smoothing consumption profiles, etc. Optimization may be based on a single objective or different objectives to be achieved simultaneously (multiobjective optimization).

3.4.1. *The architectures and components of DSM platforms*

DSM platforms are designed so as to optimally manage energy resources through a specific architecture, illustrated in Figure 3.2.

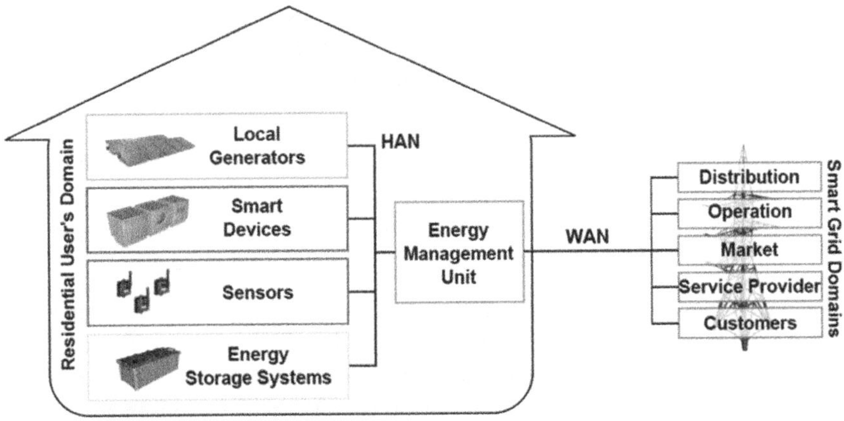

Figure 3.2. *Architecture of a DSM platform. For a color version of this figure, see www.iste.co.uk/mbarek/service.zip*

The basic components of this platform are as follows:

– local generators: these are renewable or conventional sources of energy used locally or injected into the main grid;

– smart devices: these are electrical devices that can control themselves and provide data such as their energy consumption, for instance. They may also be remotely monitored;

– sensors: these are used to collect data that are essential for the operation of a DSM system, such as position, temperature and light intensity;

– energy storage systems: these are devices that can store energy so as to allow a DSM system flexible management of energy resources;

– energy management unit (EMU): this unit is responsible for information exchange with other elements in the system and for managing users' resources, based on a smart DSM mechanism. This mechanism must determine the scheduling for devices, the operating plan for storage systems and energy demand-and-supply profiles. For multiuser architecture, all EMUs are connected to a central server that coordinates consumers;

– Smart Grid domains: these are the distribution, operation, market and supplier domains. A utility company, which is part of the market domain, supplies energy to consumers, who pay based on predefined rates. There may be an independent, not-for-profit system operator who can act as an intermediary between suppliers and consumers. This operator procures energy from the suppliers and sells it to consumers so as to maintain a balance between supply and demand.

All these components are connected through the communication infrastructure described in section 3.2.2.

3.4.2. *Classifying DSM approaches*

It is possible to classify methods to optimize DSM approaches based on three criteria (see Figure 3.3). The first criterion is the manner in which consumers are served: individually or as consumer cooperatives. In the second case, users collaborate to determine their consumption plans and DSM methods are used to optimize a shared utility function. The third criterion for classification is based on the technique used to process system parameters: deterministic or stochastic. Many DSM system parameters, the energy produced by distributed sources, consumer preferences and electricity prices, to name a few, are estimated using predictive methods. With regard to deterministic DSM approaches, these parameters are defined as deterministic data, while for the stochastic approaches they are represented by random variables, in order to include uncertainty in the decision-making process. To this end, two main methods are used: stochastic optimization and robust optimization. The first method is used for problems where uncertainty can be represented by a probabilistic model. The second method is applied if the uncertainty model is not probabilistic, if the distribution is not precisely known, or when uncertain data may be represented by a set of values. Stochastic optimization remains limited by its independence from the availability of historical data, as well as its inability to handle risk directly, which leads to several fields of applications being excluded from this method. On the other hand, robust optimization is simpler and more efficient, in practice, since it does not increase the complexity of the corresponding optimization problem. For these reasons, most of the DSM methods designed for processing uncertain data are based on robust optimization.

The final classification criterion is the time scale used for demand management: days in advance or real time. In the first case, the users' consumption plans for a future period (typically the following day) are determined *a priori*. DSM mechanisms that use this type of scheduling require forecasting certain parameters (such as the quantity of energy produced locally, electricity prices or consumer preferences) for the following day. These data may be determined by learning algorithms run by the EMU using data from the sensors, by smart devices or by external sources (for example, the weather forecast). For real-time planning, consumption plans are re-determined based on real-time events and data. Consequently, DSM systems behave in the same way as DR platforms. Real-time DSM methods are typically based on stochastic techniques.

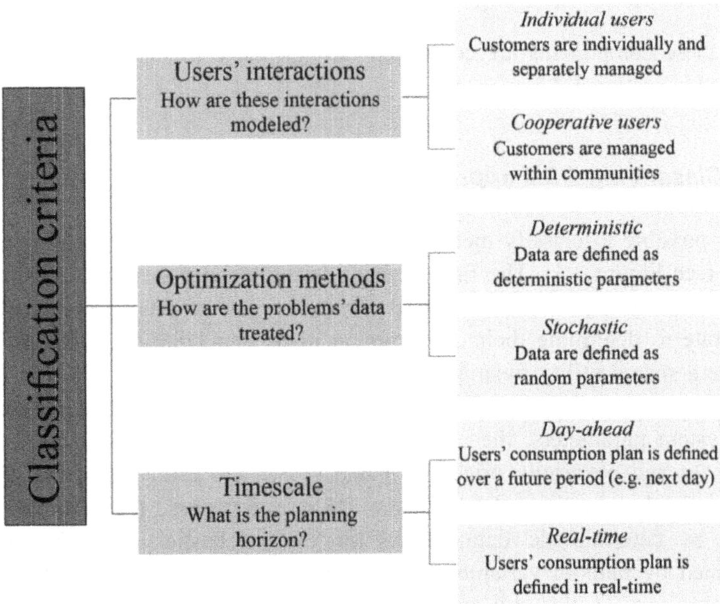

Figure 3.3. *Classification of DSM approaches*

3.4.3. *Deterministic approaches for individual users*

In this type of approach, all parameters, even those estimated by predictive algorithms, are defined as deterministic data. Ha *et al.* (2006) introduce an optimization model to minimize the thermal discomfort of the consumer through the correct use of appropriate devices. The authors use a multilevel graph in which each level is associated with a time slot and each node represents a possible state of the internal temperature. Considering this graph, the original planning problem can be solved by applying the Bellman-Ford algorithm. Goudarzi *et al.* (2011) defined a multiobjective optimization model to include the cost of consumption with thermal discomfort. They introduced two types of pricing: fixed pricing and pricing independent of consumption. They proposed an exact branch-and-bound algorithm to solve the problem. They then defined a simple but effective heuristic to avoid the complexity of the exact algorithm.

In Clastres *et al.* (2010), the planning problem was formalized as a real-time problem to adapt to the uncertainty of renewable energy resources. The authors' objective was to adjust the scheduling made in advance based on the change in energy supply. To this end, the objective function is slightly modified by the introduction of a balancing payment, a cost incurred by the consumer if the actual energy injected into the network differs from the energy defined in the pre-established schedule. Renewable resources and storage systems are also discussed in Agnetis *et al.* (2013). This project used reward tariffs and the objective function considered three parameters: cost, scheduling preferences and thermal comfort. The authors proposed a heuristic that combined two techniques, local search and exhaustive enumeration, to find suboptimal solutions in a relatively short runtime.

3.4.4. *Stochastic approaches for individual users*

In general, the generation of renewable energy, as well as the parameters for electrical devices, and sometimes, energy tariffs, is estimated by forecasting methods, which could produce errors. Consequently, appropriate tools must be used in decision-making processes for consumption scheduling when uncertainties are present. In Jacomino and Le (2012), a multiobjective method was designed to minimize the cost of consumption and the discomfort of the user. Two approaches were proposed to deal with the uncertainty of the data: the first was dedicated to the uncertainty of weather forecasting and the second considered the uncertain device data.

The uncertainty of electricity prices has been discussed in Deng *et al.* (2014). The goal of the multiobjective optimization model presented was to maximize user comfort and minimize cost, which is determined according to a fixed price (independent of demand), modeled as a random variable. The authors proposed a dual decomposition method and a stochastic gradient method. Another research project proposed a real-time system based on stochastic optimization to minimize the average cost of electricity in the long-term (Guo *et al.* 2012). The model that was introduced takes into account prices, renewable resources and storage systems. The authors used a modified version of the Lyapunov method, which was initially developed to dynamically monitor queue systems. To adapt to the problem of planning energy demand, the activities of electrical resources are represented by virtual queues. The Lyapunov model then provides solutions that are very close to the optimal solution, without the need for any statistical knowledge on the uncertain data.

The approaches that have been discussed so far are all designed to monitor the consumers' energy resources, individually and locally. However, this technique may have undesirable effects on the network, since the users' decisions are taken in a selfish manner. For example, in the case of minimizing the billable amount, all consumers may shift their demands to those periods where the prices are lowest, which would lead to peaks in consumption. To overcome these shortcomings, DSM approaches that consider consumers as communities were developed.

3.4.5. *Deterministic approaches for consumer communities*

The optimization methods used for individual users can be extended and adapted to multiuser cases. Barbato *et al.* (2011) presents a linear problem for managing the demand of a group of consumers. The main objective of this research was to minimize the total bill for consumption by the community. A centralized architecture was used, in which a central DSM controller determined the users' consumption schedules. Another centralized solution focusing on managing storage systems was proposed in Adika and Wang (2014).

Although centralized approaches of demand scheduling for consumer communities have proven to be effective, a major challenge may arise when using these approaches, related to the quantity of sensitive data that must be transmitted through the smart grid. This raises questions around scalability, security and the protection of privacy. One way of dealing with these problems consisted of designing distributed solutions for demand scheduling. Samadi *et al.* (2010) introduced a multiobjective model to monitor user devices optimally, with the goal of maximizing comfort and minimizing cost. The optimization problem is modeled in the form of a concave problem, which can be solved using convex programming techniques. The distributed approach proposed for the resolution of this problem is based on dual decomposition and the projected gradient; in particular, it uses the Lagrange duality. The solution proposed in Atzeni *et al.* (2013a) considers distributed energy generation and storage optimization. The problem is formalized in the form of a nonlinear optimization problem with which consumers determine their generation/storage strategies that allow them to minimize the aggregated cumulative cost.

3.4.6. *Stochastic approaches for consumer communities*

Some parameters may be affected by errors due to predictive algorithms. Tarasak (2011) extended the approach proposed in Samadi *et al.* (2010) to consider the uncertainty of the data, and introduced three uncertainty models: the limited uncertainty model, the Gaussian model and the unknown distribution model. The uncertainty of consumer demand was also addressed in Chang *et al.* (2013), where a real-time approach was proposed. In this research, the authors focused on plannable devices that had several operating modes. They characterized each device by two probability distributions to describe the probability of using the device in a future period and the probability of choosing a specific operating mode. The scheduling problem was formalized in the form of a multistage stochastic problem whose objective was to minimize the overall costs of the market in real-time. The Lagrange duality was used to obtain a set of subproblems, which could be solved separately and individually by each consumer, based on a Markov decision process. Jiang and Low (2011) also used the Lagrange duality and the subgradient algorithm to handle the uncertainty of the supply of renewable energy. The goal of the proposed approach was to maximize a social protection function, which was defined as being the consumer's comfort minus the cost of demand. The distinguishing characteristic of the problem presented in this paper is that it targets both advance planning and real-time planning, in order to improve overall performance.

<table>
<tr><th>Approach</th><th>User</th><th>Optimization</th><th>Goal(s)</th><th>Tools for solving</th></tr>
<tr><td>Ha et al. (2006)</td><td rowspan="6">Individual</td><td rowspan="4">Deterministic</td><td>Minimizing thermal discomfort</td><td>Bellman-Ford</td></tr>
<tr><td>Goudarzi et al. (2011)</td><td>Minimizing discomfort and cost</td><td>Branch-and-Bound</td></tr>
<tr><td>Clastres et al. (2010)</td><td>Optimizing cost and demand for renewable energy</td><td>–</td></tr>
<tr><td>Agnetis et al. (2013)</td><td>Optimizing thermal comfort, consumer preferences and cost</td><td>Local search; exhaustive enumeration</td></tr>
<tr><td>Jacomino and Le (2012)</td><td rowspan="2">Stochastic</td><td>Minimizing discomfort and cost</td><td>–</td></tr>
<tr><td>Deng et al. (2014)</td><td>Optimizing comfort and cost</td><td>Dual decomposition</td></tr>
</table>

<table>
<tr><th>Approach</th><th>User</th><th>Optimization</th><th>Goal(s)</th><th>Tools for solving</th></tr>
<tr><td>Guo et al. (2012)</td><td>Individual</td><td>Stochastic</td><td>Optimizing cost with the use of renewable energy</td><td>Lyapunov method</td></tr>
<tr><td>Barbato et al. (2011)</td><td rowspan="8">Community</td><td rowspan="4">Deterministic</td><td>Minimizing the total bill</td><td>Linear system</td></tr>
<tr><td>Adika and Wang (2014)</td><td>Optimizing the storage system</td><td></td></tr>
<tr><td>Samadi et al. (2010)</td><td>Optimizing comfort and cost</td><td>Dual decomposition; projected gradient</td></tr>
<tr><td>Atzeni et al. (2013a)</td><td>Optimizing the generation and storage of local energy</td><td>Nonlinear method</td></tr>
<tr><td>Tarasak (2011)</td><td rowspan="3">Stochastic</td><td>Optimizing comfort and cost</td><td>Uncertainty models</td></tr>
<tr><td>Chang et al. (2013)</td><td>Minimizing overall real-time costs</td><td>Lagrange dualit; Markov decision-making process</td></tr>
<tr><td>Jiang and Low (2011)</td><td>Maximizing comfort minus cost</td><td>Lagrange duality; subgradient</td></tr>
</table>

Table 3.1. *Table summarizing DSM approaches*

3.5. Techniques and methods for demand scheduling

Researchers have chosen different techniques and methods to solve the demand scheduling problem. These include three techniques that have proven to be relevant for demand management: game theory, MAS and machine learning. Scheduling approaches often use these three techniques simultaneously to combine their benefits and make the best use of them.

3.5.1. *Game theory*

Game theory is one of the most widely used techniques to develop approaches to scheduling consumer demand. This technique can express the interactions between users and the competition/cooperation between them to better manage their consumption. In Soliman and Leon-Garcia (2014), the authors proposed an approach based on a non-cooperative game to describe the competition between users, in order to find the optimal scheduling of demand to minimize consumption costs. Energy storage units are also used to reduce demand during peak hours. Consumers can thus charge the batteries outside of these hours to use or sell this energy later. However, if more users follow this technique, a new consumption peak may be produced. To solve this problem, the authors introduced a Stackelberg game, played between consumers and the supplier, in which the latter can act on the price of electricity in order to regulate demand. The major disadvantage of this approach is that users must share information on their hourly requests, which limits the practical implementation of this solution and raises the question of protecting consumers' privacy.

The authors of Atzeni *et al.* (2013b) introduced a non-cooperative game to represent the problem of managing demand in a Smart Grid with conventional (passive) consumers, as well as consumers with storage units and/or distributed sources of energy production (active consumers). The proposed approach aims to determine the optimal strategies for energy storage and production in order to optimize demand. The strategy of each active consumer depends on three parameters: their devices, the strategies of other consumers and the total consumption of passive users. The authors have shown that the Nash equilibrium can be achieved based on proximal decomposition.

Ibars *et al.* (2010) proposed a solution based on a network congestion game to guarantee that the local solution of each selfish consumer is the solution of the global optimization problem. The authors monitored energy demand during peak hours through a dynamic pricing strategy. In Mohsenian-Rad *et al.* (2010), the authors used game theory to formalize a demand scheduling game between consumers. They proposed an autonomous and distributed management system using the communication infrastructure integrated into the microgrid. Dehghanpour and Nehrir (2017) modeled the problem of demand management as a negotiation game between different entities with different objectives. They applied a Distributed Gradient Algorithm (DGA) to solve this problem in real time. In Ininahazwe *et al.* (2018), the authors proposed an autonomous distributive solution for demand-side management, based on a Bayesian game. In this game, the scheduled loads of each consumer are modeled as a three-player game where one user plays against the storage unit, and an opponent representing all the other users.

3.5.2. *Multiagent systems*

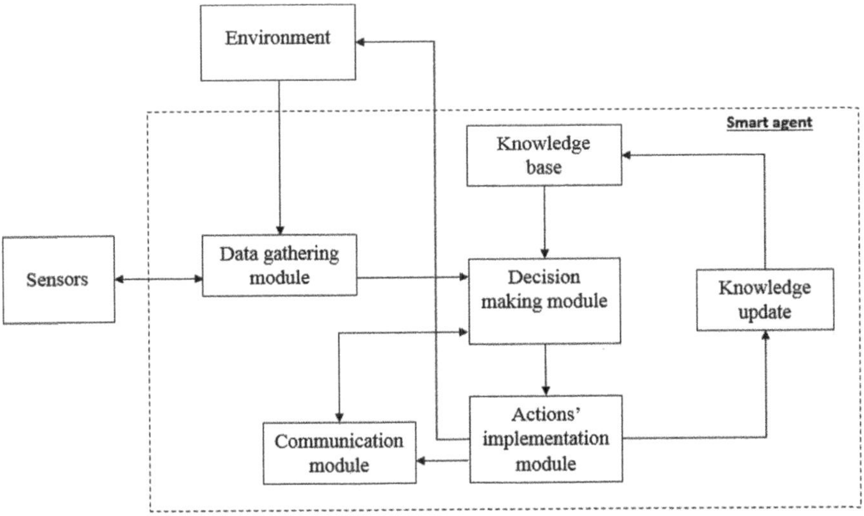

Figure 3.4. *Architecture of a smart agent*

MAS are a set of autonomous IT entities. The agents in Figure 3.4 have the ability to perceive aspects of their environment and, in many cases, to act on the environment to change it. Agents can have different degrees of intelligence depending on their roles in the MAS architecture. They can communicate with each other to exchange information or, in more complex cases, negotiate or cooperate to achieve the same objective. In the context of smart grids, MAS technology can be applied in a wide range of applications. Demand management is an essential service for smart grids, where MAS have demonstrated what they can contribute and how well they can perform. An MAS is capable of making intelligent decisions without human intervention, which makes demand management systems autonomous, flexible, scalable and fault tolerant. Consequently, many researchers have opted for MAS to develop high-performing DSM approaches.

In Anvari-Moghaddam *et al.* (2017), an ontology-oriented multiagent energy management system was proposed for optimal monitoring and control of buildings and houses in a microgrid, integrating renewable resources and controllable loads. Different agents, ranging from those with a simple reflex to those integrating complex learning, were designed and implemented to achieve the optimal operating strategy. The proposed agent-based approach solved the demand management problem using cooperation and communication between decision-making agents.

Klaimi *et al.* (2015a, 2015b, 2016) have used MAS to minimize consumer bills by optimizing the charge and discharge of the smart energy storage system. In this solution, the entities in the system (producer, consumer, storage system) are represented by agents who communicate with each other. They integrate renewable resources and try to solve their intermittency problem. However, these solutions do not take into account energy losses during transport. Klaimi *et al.* (2018) extended their research in order to minimize losses due to transmission lines. Within this research, the authors defined algorithms for negotiation and cooperation between agents, in order to maximize each entity's gain and balance the supply and demand of energy.

Chouikhi *et al.* (2018a) proposed a distributed approach to demand scheduling based on a price incentive. In this solution, consumers in the same building interact in order to aggregate their demands before sending them to the energy supplier. The supplier thus sees these consumers as being a single consumer. In addition, the price of energy depends on the total demand, which encourages users to cooperate. The scheduling problem is modeled as a multiobjective optimization problem with three objectives: (i) minimizing the total cost of consumption, (ii) reducing demand during peak hours and (iii) maximizing the consumption of renewable energy. A MAS is introduced where a Smart agent is defined for each consumer and each building. These agents are responsible for demand scheduling, knowing that it is done through a cooperative game that takes place between consumer agents, with minimum information exchange in order to protect the privacy of users. However, the energy is assumed to be sufficient to meet all the demands, which does not reflect reality. Thus, this solution was extended in Chouikhi *et al.* (2018b), where the authors added a utility function, which represents the satisfaction of each consumer, based on the availability of energy and the priorities of the consuming applications. They also added a network of wireless sensors/actuators for managing and monitoring the consumption of certain devices. This network optimizes consumption by making it possible to stop/start these devices automatically, based on several factors (for example, the temperature or the presence of the user in the apartment). In addition, a new agent is added to represent the microgrid. This agent insulates the microgrid and protects it from any disturbance in the main electrical network.

3.5.3. *Machine learning*

Machine learning focuses on designing and developing algorithms that allow control systems to predict the behavior of components in a microgrid, based on empirical data, such as the information collected by the sensors. O'Neill *et al.* (2010)

used an online learning application to implicitly assess the impact of future energy prices and consumer decisions on long-term energy costs, and thus plan the use of residential electrical devices. Fang *et al.* (2011) used online machine learning to analyze strategies for using renewable energy resources in an isolated microgrid. More specifically, the consumer tries to decide which source, among several, should be used in order to maximize their profit. Although the profile of renewable energy production is not known in advance, the authors have shown that there is a relatively small difference between the profit obtained using the optimal renewable energy source and that obtained by following their strategy.

Sheiki *et al.* (2016) proposed an automatic energy management system based on a reinforcement learning algorithm. They modeled the interaction between the electricity and natural gas networks. Their goal was to motivate residential consumers to participate in demand management programs and reduce their consumption of electricity and natural gas during peak hours.

In Kofins *et al.* (2018), the authors introduced a cooperative MAS for energy monitoring and management in an autonomous microgrid. This system learns to control the components of the microgrid, through a learning method, by distributive and collaborative reinforcement in a continuous state action space. The authors proposed the use of fuzzy Q-learning methods for agents representing the components of the microgrid, to act as independent learners who share their states in order to coordinate their behaviors. The results showed the effectiveness of individual agents in controlling the components, as well as the effectiveness of the MAS in guaranteeing the supply of electricity and increasing the reliability of the microgrid.

3.6. Conclusion

Migrating toward a smart grid allows information and communication technologies to be integrated into electrical networks, in order to better manage resources and improve the services provided. This new environment offers better management of the production, distribution and consumption of energy. However, new challenges arise and must be resolved. Researchers are therefore increasingly interested in the design and development of mechanisms and approaches that make it possible to take advantage of the features offered. It is in this context that we focused on the management of energy demand in the smart grid. This service is a promising solution for controlling consumers' electrical resources, not only in terms of reducing their bills or conserving energy, but also in improving the efficiency of the electrical network. Demand management mechanisms use techniques for consumption scheduling, adapting demand to the renewable energy generation, and responding to emergency conditions. In this chapter, we first introduced the smart grid environment and the fundamental concepts associated with it. We then examined demand management approaches, which were classified based on various criteria. We finally presented three interesting techniques and methods used in literature for energy consumption scheduling, namely game theory, MAS and machine learning. Given the importance of smart grids and the difficulties surrounding the implementation of a scalable and adaptive solution, researchers and manufacturers continue to make new contributions and, thus, continue making progress in the field.

3.7. References

Aalami, H., Moghaddem, M.P., and Yousefi, G. (2010). Demand response modeling considering interruptible/curtailable loads and capacity market programs. *Applied Energy*, 87(11), 243–250.

Adika, C.O. and Wang, L. (2014). Smart charging and appliance scheduling approaches to demand side management. *International Journal of Electrical Power and Energy Systems*, 57, 232–240.

Agnetis, A., De Pascale, G., Detti, P., and Vicino, A. (2013). Load scheduling for household energy consumption optimization. *IEEE Transactions on Smart Grid*, 4(14), 2364–2373.

Akyildiz, I.F., Su, W., Sankarasubramaniam, Y., and Cayerci, E. (2002). Wireless sensor networks: A survey. *Computer Networks*, 38(14), 393–422.

Andrews, J.G., Ghosh, A., and Muhamed, R. (2007). *Fundamentals of WiMAX: Understanding Broadband Wireless Networking*. Pearson Education, Upper Saddle River, New Jersey, USA.

Anvari-Moghaddam, A., Rahimi-Kian, A., Mirian, M.S., and Guerrero, J.M. (2017). A multi-agent based energy management solution for integrated buildings and microgrid system. *Applied Energy*, 203, 41–56.

Atzeni, I., Ordonez, L.G., Scutari, G., Palomar, D.P., and Fonollosa, J.R. (2013a). Noncooperative and cooperative optimization of distributed energy generation and storage in the demand-side of the smart grid. *IEEE Transactions on Signal Processing*, 61(11), 2454–2472.

Atzeni, I., Ordonez, L.G., Scutari, G., Palomar, D.P., and Fonollosa, J.R. (2013b). Demand-side management via distributed energy generation and storage optimization. *IEEE Transactions on Smart Grid*, 4(12), 866–876.

Barbato, A., Capone, A., Carello, G., Delfanti, M., Merlo, M., and Zaminga, A. (2011). House energy demand optimization in single and multi-user scenarios. *IEEE International Conference on Smart Grid Communications (SmartGrid Comm)*. IEEE, Brussels, Belgium.

Chang, T.H., Alizadeh, M., and Scaglione, A. (2013). Real-time power balancing via decentralized coordinated home energy scheduling. *IEEE Transactions on Smart Grid*, 4(13), 1490–1504.

Chouikhi, S., Merghem-Boulahia, L., and Esseghir, M. (2018a). Energy demand scheduling based on game theory for microgrids. *IEEE International Conference on Communications (ICC)*. IEEE, Kansas City, USA.

Chouikhi, S., Merghem-Boulahia, L., and Esseghir, M. (2018b). Multi-level energy consumption optimization for smart buildings. *IEEE Global Communications Conference (Globecom)*. IEEE, Abu Dhabi, United Arab Emirates.

Clastres, C., Ha Pham, T., Wurtz, F., and Bacha, S. (2010). Ancillary services and optimal household energy management with photovoltaic production. *Energy*, 35, 55–64.

Coelho, V.N., Coelho, I.M., Coelho, B.N., Reis, J.R., Enayatiar, R., Souza, M.J.F., and Guimarães, F.G. (2016). Self-adaptive evolutionary fuzzy model for load forecasting problems on smart grid environment. *Applied Energy*, 169, 567–584.

Dehghanpour, K. and Nehrir, H. (2017). Real-time multiobjective microgrid power management using distributed optimization in an agent-based bargaining framework. *IEEE Transactions on Smart Grid*, 6(9), 6318–6327.

Deng, R., Yang, Z., Chan, J., and Chow, M. (2014). Load scheduling with price uncertainty and temporally-coupled constraints in smart grids. *IEEE Transactions on Power Systems*, 29(16), 2823–2834.

Fang, X., Misra, S., Xue, G., and Yang, D. (2012). Smart Grid – The new and improved power grid: A survey. *IEEE Communications Surveys & Tutorials*, 14, 944–980.

Fang, X., Yang, D., and Xue, G. (2011). Online strategizing distributed renewable energy resource access in islanded microgrids. *IEEE Global Telecommunications Conference (GLOBECOM)*. IEEE, Houston, USA.

Ghasemi, A., Shayeghi, H., Moradzadeh, M., and Nooshyar, M. (2016). A novel hybrid algorithm for electricity price and load forecasting in smart grids with demand-side management. *Applied Energy*, 177, 40–59.

Goudarzi, H., Hatami, S., and Pedram, M. (2011). Demand-side load scheduling incentivized by dynamic energy prices. *IEEE International Conference on Smart Grid Communications (SmartGridComm)*. IEEE, Brussels, Belgium.

Guo, Y., Pan, M., and Fang, Y. (2012). Optimal power management of residential customers in the smart grid. *IEEE Transactions on Parallel and Distributed Systems*, 23(19), 1593–1606.

Ha, D.L., Ploix, S., Zamai, E., and Jacomino, M. (2006). A home automation system to improve household energy control. *IFAC Proceedings*, 39(13), 15–20.

Hiertz, G.R., Denteneer, D., Stibo, L., Zang, Y., Costa, X.P.C., and Walke, B. (2010). The IEEE 802.11 universe. *IEEE Communications Magazine*, 48(11), 62–70.

Holyhead, J., Ramchurn, S., and Rogers, A. (2015). Consumer targeting in residential demand response programmes. *ACM Sixth International Conference on Future Energy Systems*. ACM, Bangalore, India.

Ibars, C., Navarro, M., and Giupponi, L. (2010). Distributed demand management in smart grid with a congestion game. *IEEE International Conference on Smart Grid Communications (SmartGridComm)*. IEEE, Gaithersburg, USA.

Ininahazwe, H., Muriithi, C.M., and Kamau, S. (2018). Optimal demand-side management for smart micro grid with storage. *Journal of Power and Energy Engineering*, 6(12), 38–58.

Jacomino, M. and Le, M.H. (2012). Robust energy planning in buildings with energy and comfort costs. *4OR*, 10, 81–103.

Jiang, L. and Low, S. (2011). Multi-period optimal energy procurement and demand response in smart grid with uncertain supply. *IEEE Conference on Decision and Control and European Control Conference (CDC-ECC)*. IEEE, Orlando, USA.

Kirby, B. and Hirst, E. (1997). Ancillary service details: Voltage control. Report, Oak Ridge National Laboratory, Oak Ridge, USA.

Klaimi, J., Merghem-Boulahia, L., Rahim-Amoud, R., and Jrad, A. (2015a). An energy management approach for smart-grids using intelligent storage systems. *International Conference on Digital Information and Communication Technology and Its Applications (DICTAP)*. IEE, Beirut, Lebanon.

Klaimi, J., Rahim-Amoud, R., Merghem-Boulahia, L., and Jrad, A. (2015b). An agent-based approach for energy management in smart-grids. In *Highlights of Practical Applications of Agents, Multi-agent Systems, and Sustainability – The PAAMS Collection*, Bajo, J., Hallenborg, K., Pawlewski, P., Sánchez-Pi, N., Duque Méndez, N.D., Lopes, F., and Julian, V. (eds). Springer, Berlin, Germany.

Klaimi, J., Rahim-Amoud, R., Merghem-Boulahia, L., and Jrad, A. (2018). A novel loss-based energy management approach for smart grids using multi-agent systems and intelligent storage systems. *Sustainable Cities and Society*, 39, 344–357.

Klaimi, J., Rahim-Amoud, R., Merghem-Boulahia, L., Jrad, A., and Esseghir, M. (2016). An intelligent storage-based energy management approach for smart grids. *International Conference on Wireless and Mobile Computing, Networking and Communications (WiMob)*. IEEE, New York, USA.

Kofins, P., Dounis, A., and Vouros, G. (2018). Fuzzy Q-learning for multi-agent decentralized energy management in microgrids. *Applied Energy*, 219, 53–67.

Mohsenian-Rad, A.H., Wong, V.W., Jatskevich, J., Schober, R., and Leon-Garcia, A. (2010). Autonomous demand-side management based on game theoric energy consumption scheduling for the future smart grid. *IEEE Transactions on Smart Grid*, 1(13), 320–331.

O'Neill, D., Levorato, M., Goldsmith, A. and Mitra, U. (2010). Residential demand response using reinforcement learning. *IEEE International Conference on Smart Grid Communications*. IEEE, Gaithersburg, USA.

Olsson, M., Sultana, S., Rommer, S., Frid, L., and Mulligan, C. (2009). *SAE and the Evolved Packet Core: Driving the Mobile Broadband Revolution*. Elsevier, New York, USA.

Saebi, J., Taheri, H., Mohammadi, J., and Nayer, S.S. (2010). Demand bidding/buyback modeling and its impact on market clearing price. *IEEE International Energy Conference and Exhibition (EnergyCon)*. IEEE, Manama, Bahrain.

Sahebi, M.M., Duki, E.A., Kia, M., Soroudi, A., and Ehsan, M. (2012). Simultaneous emergency demand response programming and unit commitment programming in comparison with interruptible load contracts. *IET Generation Transmission & Distribution*, 6(17), 605–611.

Samadi, P., Mohsenian-Rad, A.-H., Schober, R., Wong, V.W.S., and Jatskevich, J. (2010). Optimal real-time pricing algorithm based on utility maximization for smart grid. *IEEE International Conference on Smart Grid Communications (SmartGridComm)*. IEEE, Gaithersburg, USA.

Sesia, S., Toufik, I., and Baker, M. (2011). *LTE – The UMTS Long Term Evolution*. John Wiley & Sons, West Sussex, UK.

Sheikhi, A., Rayati, M., and Ranjbar, A.M. (2016). Demand side management for a residential customer in multi-energy systems. *Sustainable Cities and Society*, 22, 63–77.

Soliman, H.M. and Leon-Garcia, A. (2014). Game-theoretic demand-side management with storage devices for the future smart grid. *IEEE Transactions on Smart Grid*, 5(13), 1475–1485.

Tarasak, P. (2011). Optimal real-time pricing under load uncertainty based on utility maximization for smart grid. *IEEE International Conference on Smart Grid Communications (SmartGridComm).* IEEE, Brussels, Belgium.

Trehan, N. (2001). Ancillary services-reactive and voltage control. *IEEE Power Engineering Society Winter Meeting*. IEEE, Columbus, USA.

Xi, F., Satyajayant, M., Guoliang, X., and Dejun, Y. (2012). Smart Grid – The new and improved power grid: A survey. *IEEE Communications Surveys & Tutorials*, 14, 944–980.

Zhu, L., Yan, Z., Lee, W.J., Yang, X., Fu, Y., and Cao, W. (2015). Direct load control in microgrids to enhance the performance of integrated resources planning. *IEEE Transactions on Industry Applications*, 51(5), 3553–3560.

4

Managing Quality of Service and Security in an e-Health Environment

Mohamed-Aymen CHALOUF
IRISA, University of Rennes 1, Lannion, France

4.1. Introduction

Over the past few years, there has been a dramatic increase in the number of aged persons and patients with limited autonomy or chronic diseases. Providing healthcare to these individuals, whether in retirement homes or in hospitals, is a very demanding service.

As a result of these factors, a large number of national and international projects have been developed over recent years. These projects have a common objective, namely, ensuring that individuals with reduced autonomy or chronic diseases become able to stay at home, while providing them with necessary aid and essential healthcare.

The solutions proposed in studies carried out in this field use information and communication technologies to enable health professionals (doctors and nurses) to remotely monitor the state of health of the concerned person. These solutions also allow these individuals to access their health data such as diagnostics, for instance, and to get help when needed.

Service Level Management in Emerging Environments,
coordinated by Nader MBAREK.

The integration of Information and Communication Technologies (ICTs) into the medical field has, among other things, promoted innovation in diagnostics and in tracking patients and also given rise to the world of "e-health". This is defined as the "use of ICT to provide medical services". e-health requires multidisciplinary technologies such as instrumentation, information technology and communication technology for the collection, transmission, processing and storage of medical information and for guaranteeing medical services remotely. Thus, e-health designates the use of ICT for all healthcare activities and the concept integrates telemedicine, tele-health as well as other digital healthcare solutions.

Telemedicine primarily covers four activities (Agence Française de la Santé Numérique 2010; Simon 2010):

– teleconsultation: this allows a consultation to be carried out remotely, with access to the patient's medical files and data. In this case, the patient is brought into contact with a healthcare professional, preferably equipped with a camera (e.g. a webcam or smartphone camera) so as to allow identification;

– teleexpertise: this allows doctors to remotely communicate and exchange thoughts and medical opinion based on a patient's medical file. A doctor may ask their colleagues for their opinions on how best to treat a patient by sharing information from the patient's medical files, such as images (e.g. from a scan or radiography);

– telecare: this allows a patient suffering a chronic disease to monitor certain aspects of their health at home and to easily share this medical information with healthcare professionals. These professionals can, thus, remotely monitor the results and adjust treatment accordingly;

– telemonitoring: this mechanism allows the remote medical monitoring of a person suffering from a serious chronic disease or of an elderly person with limited autonomy. It facilitates the tracking of the concerned person by guaranteeing a more efficient remote data processing. The data are obtained by monitoring certain biological constants such as oximetry and cardiac rhythm. Depending on the person's profile, this monitoring may be continuous or discontinuous.

A telemedicine service may be carried out synchronously or asynchronously. Synchronous telemedicine requires that both parties be present at the same time and that they interact through a communication means such as videoconferencing. In asynchronous telemedicine, both parties do not need to be present at the same time. Indeed, the patient's medical data may be shared with a health professional (e.g. a doctor) who will consult the data later. If we take the example of telemonitoring, then this telemedicine service may be carried out synchronously or asynchronously (Moulin and Simon 2016). Asynchronous telemonitoring, for instance, is used for patients with a stabilized chronic disease. In this case, the healthcare professional requires information on events that occur over a certain period (e.g. the consultations are a few months apart). This information may be with respect to the stability of the patient's weight and their blood pressure or blood sugar readings. Asynchronous telemonitoring is interesting as it can reduce the complications inherent to a chronic disease (e.g. accidents linked to an ongoing treatment). Synchronous telemonitoring, on the other hand, is used for patients suffering from a severe disease with the risk of complications. For example, a patient suffering advanced cardiac troubles with the risk of complication and emergency hospitalization must be telemonitored in real time. This makes it possible to have an immediate response in the case of any alerts in order to avoid complications and hospitalization. To summarize, the modalities of telemonitoring (synchronous or asynchronous) must be defined based on the patient's profile. It must be noted that in certain cases periods of asynchronous telemonitoring can alternate with periods of synchronous telemonitoring.

In section 4.2, we describe e-health systems through the examples of some known European projects in this field. These were the earliest European projects that studied different patient profiles (Hamdi *et al.* 2014). As several studies, implementations and simulations have shown, e-health systems need to manage Quality of Service (QoS) and security in order to respond to the requirements of the e-health services and to be able to resist malicious attacks. We will, therefore, discuss the QoS control and security management in sections 4.3 and 4.4, respectively. The chapter then concludes with section 4.5.

4.2. e-health systems

Our survey on the most representative European research projects in the e-health domain (Hamdi *et al.* 2014) allowed us to identify the architecture that is common to e-health systems and to identify its main characteristics. The e-health systems studied in Hamdi *et al.* (2014) mainly focused on telemonitoring and make it possible to provide services to the monitored persons: the elderly (Rocha *et al.* 2011; *SOPRANO Project* 2011), the disabled (Navarro *et al.* 2011) and patients suffering chronic diseases (Chronious Project 2012; REACTION Project 2014). These services basically cover safety services such as detecting gas leaks, fires or an intrusion, as well as health services such as reminders for medication, remote diagnosis and detecting dangerous situations. The e-health service is implemented using Open Services Gateway initiative (OSGi) (*SOPRANO Project* 2011) or ontologies (Chronious Project 2012; REACTION Project 2014). It must also be noted that the implementation of health services is often distributed between local equipment (generally a gateway located in the residence of the monitored person) and remote equipment (generally distant servers, such as Cloud servers). This makes it possible to have "local" services and make local entities as autonomous as possible.

4.2.1. *Architecture*

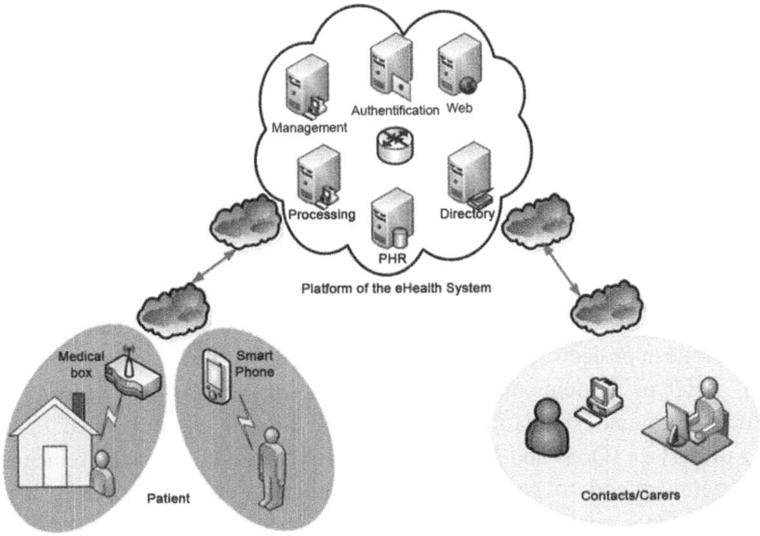

Figure 4.1. *The architecture of e-health systems (Hamdi* et al. *2014)*

The "common" architecture is an architecture that may represent the point of convergence of the e-health solutions proposed in the studied projects. This architecture is composed of three main parts (see Figure 4.1):

– sensor network: composed of various sensors (medical, environmental, contextual, etc.) equipping the patient and their environment. Each provided service is based on a set of sensors. For example, remote diagnosis requires medical sensors that measure physiological parameters such as body temperature and pulse oximetry. However, a service that detects falls may use localization and activity sensors. The same sensor may also be involved in several different health services;

– gateway: this is intermediate equipment between the patient in their environment and the outside world. When the telemonitored person is at home, this gateway is a sort of medical box. However, a smartphone or PDA can replace the medical box to provide continuous telemonitoring when the person is away from home. The gateway is also the equipment that runs local services and, generally, the local functionalities of a service (an e-health service may be made up of local and remote functionalities);

– remote servers: these servers are connected to the gateway, which must transmit data about the patient and their environment to these servers. They can carry out complex computations and store a large amount of data. Remote servers are thus used to run remote services and, generally, remote functionalities for any service. In some cases, these servers can decide whether or not to alert a third person about the state of the monitored person (in the case of an incident, for instance). Remote servers therefore have access to a contact list (a parent, a neighbor, a nurse, a doctor, emergency services, etc.).

To support communication between the sensors and the gateway, the latter must support various communication technologies (e.g. Ethernet, Wi-Fi, Bluetooth, Radio Frequency, etc.). This allows the gateway to communicate with remote servers as well as the various sensors equipping the patient and their environment. Communications between the gateway and the remote servers are often based on web services in order to ensure interoperability (REACTION Project 2014).

4.2.2. *Characteristics*

This section describes the principal characteristics of an e-health system: the services that an e-health system must provide in order to satisfy user needs (e.g. a patient or elderly person), the technical tools and technologies used to guarantee these services, and the supported functionalities. We will then present some technical problems that may be encountered.

4.2.2.1. *Services offered*

The main services that may be offered to users of an e-health system are as follows:

– safety: the system must guarantee the safety of the monitored person and the security of their environment by installing sensors that can detect falls, fires, etc. This allows the user to not only have a clear visibility of what is going on in their environment but also gives them the ability to control it;

– localization: the system must be able to provide the user's geographical position and location if it is needed. This makes it possible, for instance, to detect dangerous situations or send rescue to the right place;

– social interaction: the e-health system must be able to guarantee communication, information, maintaining contact and active participation in society;

– real-time alert: the e-health system may be responsible for the surveillance of the state of health of a person presenting serious risks. In this case, time is a crucial factor and as soon as a problem is detected the system must be able to react instantly and transmit real-time alerts to the concerned healthcare professional and possibly other contacts (a relative or a neighbor, for instance);

– support and coaching: a telemonitoring system must allow healthcare professional to stay in contact with the users and access all necessary data at any time and everywhere. This will guarantee efficient supervision.

4.2.2.2. *Tools and technologies*

The most used technical tools used to implement a telemonitoring system are as follows (Hamdi *et al.* 2014):

– communication techniques: the architecture presented in section 4.2.1 shows various types of communication within e-health systems – the communications between the sensors equipping the patient and their environment as well as communications between the sensors and the gateway. These two types of communication may be wired or wireless (Wi-Fi, Bluetooth, etc.). We then have the communications between the gateway and the remote servers. These communications are generally over the Internet using a wired connection (ADSL, cable, optical fiber) or a mobile connection (2G, 3G, UMTS, 4G). Finally, we have the communications between the servers of the e-health system platform and the terminals of the healthcare professionals or any other contact associated with the monitored person. These communications may use the Internet or a mobile network (an SMS alert, for example);

– computation and processing: the level of processing and computation carried out in the context of the e-health service differ from one project to another as they are related to the profile of the person using the e-health system. However, we have identified certain criteria that made it possible to differentiate between e-health systems. The first criterion concerns the nature of the processing that is carried out on the medical data: distributed or centralized. Another criterion that could be used to characterize e-health systems is the degree of reactivity of this system. Indeed, in the majority of cases, the e-health system must be able to react in real time when an incident occurs;

– software: an e-health system must implement a software application that allows users to easily access the available services. In this case, great importance must be given to the interface that allows users to interact with the e-health system;

– electronic health record (EHR): this allows the storage of health data as well as any other relevant information about the user;

– user profile: when used in e-health systems, the user profile gathers all information concerning the user and their preferences. This information is taken into account by the e-health system when running e-health services, which makes it easier to offer user-centered services.

4.2.2.3. *Specific criteria*

The main criteria that characterize an e-health system are as follows:

– mobility: e-health systems today aim to provide services such as telemonitoring whether the person is near their house or far from home. Technologically advanced mobile devices equipped with greater capacities (larger memory, greater computing power, longer battery life, etc.) have made it easier to implement e-health services for mobile users;

– security: the various components of an e-health system may carry out tasks related to the processing, storage, transmission and accessing of critical medical data. These systems must, therefore, be secured in order to guarantee secure storage, access and communications. This allows the system to resist attacks that could damage the confidentiality, integrity and availability of data;

– context awareness: an e-health system must be able to take into account the context and history of the telemonitored person. This allows the system to adjust certain parameters in order to be more effective in its actions and in the reports transmitted to the users and healthcare professionals;

– interoperability: the e-health system must be able to function with existing and future hardware, products and systems. It must not, therefore, restrict access or implementation.

The earliest e-health systems neglected the quality and security aspects. Indeed, very few projects considered the quality of transmissions and maintaining connectivity despite the importance of these aspects in guaranteeing a continuous and high-quality e-health service. As for security, a false alert or a blocked alert could have disastrous consequences on the user's health, which could even be fatal. As was stated above, despite its importance, security was largely neglected in the earliest e-health systems. Currently, e-health systems are well developed and have become a reality. Thus, we are finding more and more research that focuses on the security and quality aspects. In the following sections, we will describe the leading efforts that contributed to quality and security management in e-health systems.

4.3. QoS in e-health systems

In the previous sections, we gave an overview of e-health services, described the services that may be offered, their general architecture, their main characteristics and the technological tools on which they are based. We will now look at some relevant work that examines QoS management in e-health systems.

It is essential to manage QoS in e-health systems as we have e-health services with different requirements in terms of resources (e.g. bandwidth, time, jitter, packet loss rate). Further, e-health systems have a special architecture with different segments that involve different communication technologies. This can complicate the implementation of QoS management mechanisms. There are various mechanisms that are adapted to different parts of an e-health system, but the primary goal must be to ensure that the QoS provided to the users is adequate.

In this section, we will first review the requirements of e-health services in terms of quality. We will then describe the principal quality management mechanisms for different parts of an e-health system.

4.3.1. *e-health services and QoS*

e-health services include a large number of services with different requirements in terms of QoS. To be more precise, these services have different needs in terms of time, bandwidth, loss rate, etc. In terms of time, for example, an e-health service's requirements may go beyond real-time transmission (for example, transmitting an urgent alert) to delay-tolerant transmission (for example, accessing a patient's electronic health record). In terms of packet loss rate, an e-health service may offer a voice conversation and may, therefore, tolerate a certain loss of packets. However, it could also be a service for transmitting medical images and in this case it will not tolerate loss.

4.3.1.1. *Importance of QoS for e-health services*

A large number of research projects have provided a classification of e-health services based on their needs while also evaluating the network's ability to satisfy these needs. Salehi and Mirhadi (2008) classified different e-health services based on the importance of some QoS parameters. In Perakis and Koutsouris (2009), e-health services were classified based on the telemedicine procedures we discussed earlier (telediagnosis, teleconsultation, telemonitoring, etc.). Vouyioukas *et al.* (2007) proposed another classification of e-health services. This classification is based on QoS service requirements and distinguishes two main categories: real-time and near-real-time services. This classification was then improved by taking into account the overall context of the e-health service (Skorin-Kapov and Matijasevic 2010). The QoS needs for the same e-health service (e.g. telediagnosis) may depend on the overall context of the patient. For example, the telemonitoring of a patient in an emergency requires real-time transmission of medical data, whereas for the telemonitoring of a patient in a non-emergency situation, the data transmission could tolerate a delay. Given the importance of contextual information, such as the urgency of the patient's situation, the classification proposed in Vouyioukas *et al.* (2007) has been extended to include the concept of context awareness. Table 4.1 (Skorin-Kapov and Matijasevic 2010) illustrates this new classification where context awareness refers to the degree of urgency of the e-health service. Thus, all e-health services that could be used in an emergency are marked as being context sensitive (Skorin-Kapov and Matijasevic 2010).

Application	Bandwidth	Low delay	Low jitter	Context sensitive
Telediagnosis	High	Yes	No	Yes
Teleconsultation	High	Yes	Yes	Yes
Telemonitoring	Low	No	No	Yes
Access to file	High/low	No	No	Yes

Table 4.1. *Importance of QoS for e-health procedures/services (Skorin-Kapov and Matijasevic 2010)*

4.3.1.2. *QoS needs of e-health services*

E-health services include multimedia conferencing, the transmission of physiological parameters and medical data, sharing of medical images, access to the patient's electronic health record and so on. These services require the transport of various types of data that have different needs in terms of delay, bandwidth and packet loss. In this section, we will describe the needs of the main e-health services in terms of QoS.

4.3.1.2.1. Multimedia conferencing

Multimedia conferencing (audio-conference or videoconference) is used in e-health systems for different types of communication: between a patient and their doctor, between two doctors or between two patients. This involves transmitting audio and video and may tolerate packet loss in order to fulfill requirements in terms of time and providing an interactive real-time service. These requirements may depend on the context of e-health service (Skorin-Kapov and Matijasevic 2010). For instance, a service involving an audio/video conference between a patient and a doctor for a routine examination can be considered "non-critical" and can tolerate considerably poor quality in terms of time (150–400 msec). On the other hand, if the same service is used in an emergency, the network must be able to guarantee real-time transmission and the delay must be no greater than 150 msec. Furthermore, in Vouyioukas *et al.* (2007) we see a difference between the QoS needs of real-time video transmission, offline video transmission, multimedia conference for telediagnosis, etc. In these projects, real-time video transmission for telediagnosis is considered as more demanding in terms of QoS. We then look at real-time audio transmission for telediagnosis. This audio transmission may include the sound of a stethoscope or the audio stream that accompanies the telediagnosis video.

4.3.1.2.2. Transfer/streaming of medical images

The fixed medical images that can be transmitted within an e-health service (for example, teleconsultation) essentially cover (Perakis and Koutsouris 2009) X-rays, magnetic resonance images (MRI), ultrasound, etc. The transmission of these images in high definition requires a packet loss rate of almost zero. The transfer of this type of medical data requires great reliability because a loss of packets can have negative consequences on the diagnostic process. On the other hand, this e-health service has no specific bandwidth requirements. It must be noted that low bandwidth leads to longer transmission delays. The image size and bit rate required for different medical equipment and devices given in Vouyioukas *et al.* (2007) are summarized in Table 4.2.

Equipment	Time/space (samples/sec)	Contrast/resolution (bits per sample)	Required bandwidth
Digital blood pressure monitor	1	×10	<10 kb/sec
Digital stethoscope	10,000	×12	~120 kb/sec
Electrocardiogram	1,250	×12	~15 kb/sec
Radiology, ultrasound	512 × 512	×8	256 kb
Scanned X-ray	1,024 × 1,250	×12	1.8 Mb
Mammography	4,096 × 4,096	×12	24 Mb
Compressed video	–	–	384 kb/sec at 1.5 Mb/sec

Table 4.2. *Medical images: characteristics and QoS required (Vouyioukas* et al. *2007)*

4.3.1.2.3. Transmission of a patient's vital signs

In the context of an e-health service such as telemonitoring, the amount of information transmitted over the network related to the patient's vital signs, as well as the frequency of these transmissions depends essentially on the needs of the patient and their overall context (for example, chronic disease, emergency, etc.). In some cases, it is necessary to transmit vital signs every few minutes, while in other cases it is essential to transmit them every few seconds. Kumar *et al.* (2008) discuss the QoS requirements of a telemonitoring service for cardiac patients. In this context, the e-health system comprised wireless biomedical sensors (SpO_2, electrocardiogram, heart rate, blood pressure, etc.), which communicated with a module responsible for processing the signals received and their transmission via various network interfaces to remote servers. The decision made by this module about the transmission of physiological data depended on the transmission thresholds and frequencies specified by healthcare professionals. Thus, QoS requirements in the context of telemonitoring depend on the size of the data transmitted, the frequency of this transmission (bandwidth) as well as how urgent the data are (time).

4.3.1.2.4. Access to the electronic health record

The information systems in hospitals and healthcare institutions often use an EHR, which contains the different data related to a person's health. Thus, in the context of e-health service provision, different components of the e-health systems need to access and handle this EHR. The access can be established in different situations: by the person in their house or while they are moving, in an emergency situation or a non-urgent situation, etc. Regardless of the overall situation, access to medical data requires great reliability and the transmission, must, therefore take place with no loss of packets. However, the transmission time, which usually depends on the amount of transmitted information, must not exceed a certain threshold in case of an urgent situation.

An e-health service may involve simultaneous transmission of multiple flows. For example, an ambulance connected to the e-health system must transmit various types of data in real time: audio, video, medical images and physiological parameters. For example, Martinez *et al.* (2008) defined a telemonitoring system for patients treated in an ambulance connected via the 3G network.

4.3.2. *QoS management in e-health systems*

The convergence of networks (television, voice and data) has enabled the appearance of new services and new applications based on IP networks. Among these services, we have e-health services, which increase the diversity of the flows transported by the IP networks. The basic operation of the IP protocol does not guarantee QoS for specific flows such as those of e-health services. However, other mechanisms can be used to manage QoS. Thus, to meet the needs of flows related to e-health services transported by IP networks at the same time as other flows (for example, voice, video, mail, Web, etc.), mechanisms dedicated to the implementation of QoS, such as defining classes of traffic and the management of priorities as well as the reservation of resources, are essential. In Vergados (2007) and Qiao and Koutsakis (2008), mechanisms for managing priorities and allocating resources for various e-health flows were defined. In the following section, we describe the principal efforts around managing QoS in e-health systems.

4.3.2.1. *Mobile network-based e-health systems*

The earliest research into QoS management of e-health services focused on the early e-health systems where access to the IP network was often established through cellular networks (2G, 3G, etc.). This research is focused on the development of e-health systems using mobile or wireless networks to meet very specific needs (for example teleconsultation, telemonitoring of vital signs, etc.) as well as the evaluation of the capacity of these networks to transport flows related to e-health services. A large number of projects are focused on mobile networks (essentially 3G networks) and in their ability to transport e-health flows for urgent services (Chu and Ganz 2004; Gállego *et al.* 2005; Navarro *et al.* 2006; Vouyioukas *et al.* 2007; Martinez *et al.* 2008), teleconsultation services (Garawi *et al.* 2006; Perakis and Koutsouris 2009) or even telemonitoring services (Konstantas *et al.* 2004; HealthService 24 2006). In addition, other studies have looked at other wireless networks and their capacity to transport e-health flows (for example urgent medical flows) (Chu and Ganz 2007; Zvikhachevskaya *et al.* 2009).

Some of these projects showed that the networks used were of good quality and could support the transport of flows generated by the considered e-health services (Kumar *et al.* 2008). In these cases, the available resources were able to satisfy the QoS requirements of the e-health services. For example, the research described in Gállego *et al.* (2005) studied the simulation of a realistic e-health system made up of several ambulances, where each was able to transmit various data (ECG signals, medical analysis, voice and video) to a hospital using an UMTS network (Universal Mobile Telecommunications System). The QoS requirements for the transmitted flows are summarized in Table 4.3. The simulation results showed that the UMTS was able to respond perfectly to the defined needs.

Service	Flow rate	Maximum delay	Packet loss
Audio	4–25 kbit/sec	150–400 msec	3%
Video	32–384 kbit/sec	150–400 msec	1%
Electrocardiogram	1–20 kbit/sec	~ 1 sec	0
File transfer	–	–	0

Table 4.3. *QoS required for the constrained e-health flows (Gállego* et al. *2005)*

On the other hand, the e-health systems implemented within the framework of other projects have shown certain limitations. For instance, the evaluation of the telemonitoring system developed within the project MobiHealth (Konstantas *et al.* 2004) based on different services with different bandwidth needs (low: less than 12 kbit/sec, average: 12–24 kbit/sec and high: greater than 24 kbit/sec) and delay needs (periodic transmission and real-time transmission) revealed the limitations of the networks used (GPRS and UMTS). In fact, the bandwidth for uplink traffic was very limited and did not meet the needs of the e-health system, because in telemonitoring systems the data flows are higher on the user side than on the server side. In addition, significant delays ranging from 100 msec (for 174 byte packets) to 1,200 msec (for 8,122 byte packets) and connection losses have been observed.

Various solutions have been proposed in order to best respond to the needs of e-health flows transported by these systems. A large number of these solutions are based on the differentiation of flows based on their needs and the assignment of different priorities to these flows. This allows the network to adapt its behavior according to the nature of the transported flows. For example Martinez *et al.* (2008) defined a telemonitoring system for patients treated in an ambulance connected via the 3G network. To meet the different QoS requirements of the various transported flows, the authors proposed adaptive QoS decision mechanisms based on the available network resources. They identified different types of services (audio, video, medical images, physiological parameters, administrative data, etc.), which can be combined in different ways depending on the availability of resources to provide an optimal telemonitoring service. In the same way, a mobile telemonitoring system capable of transmitting different types of data simultaneously over the 3G network (video, medical images and ECG signals) was evaluated (Chu and Ganz 2007). Various flow parameters were tested: sampling rate, frequencies, resolutions, compressions, etc. This made it possible to define, in conjunction with healthcare professionals, the minimum levels of QoS that can be tolerated. Thus, three priority levels have been defined: (1) low priority for video streams (loss tolerance, delay intolerance), (2) medium priority for images (no tolerance to loss, delay tolerance) and (3) high priority for ECG signals (no tolerance to loss or to delay). Similarly, the work carried out in Qiao and Koutsakis (2008) defined priorities for e-health flows while proposing new scheduling strategies in order to allow mobile networks to meet the QoS needs of e-health services in the presence of other flows. These strategies essentially made it possible to prioritize urgent e-health flows (video, medical images, ECG signals) while satisfying the QoS needs of other flows. The results of the performance evaluation of the proposed solution were quite satisfactory. Indeed, for a flow related to e-health services representing 10% of all flows, the transmission delay as well as packet loss rate are much lower than the limits set by health professionals for the different types of e-health flows. Vouyioukas *et al.* (2007) studied the same problem. The researchers evaluated the ability of the 3G HSPA (High-Speed Packet Access) network to satisfy the QoS needs of a joint transmission of various urgent e-health flows (voice, video, vital signals and file transfer) under conditions of congestion. The results obtained showed the effectiveness of the prioritization of e-health traffic, which was the result of the changes made to the functioning of the HSDPA planner. This made it possible to transmit critical e-health flows on priority and to limit transmission delay to acceptable values (for example, 150 msec for voice and 240 msec for video). Other research projects studied the ability of the 4G mobile network to support e-health services and their specific QoS needs. This resulted in the birth of the concept of 4G-health. This was defined as the evolution of m-health to 4G mobility. In this context, Istepanian *et al.* (2012) present some challenges of 4G-Health and illustrate the importance of the concepts of medical Quality of Service (m-QoS) and medical Quality of Experience (m-QoE) for e-health systems based on the 4G network. In Istepanian *et al.* (2012), the authors also present a scenario describing a medical video broadcasting service to validate the concepts introduced.

4.3.2.2. *New Generation Network-based e-health systems*

Other research projects have explored the capacity of New Generation Networks (NGNs) to support e-health services. NGNs are characterized by a network center based on IP technology and a variety of access networks (wired, wireless, mobile etc.), and are able to transport any type of flow (television, voice and data).

The massive deployment of wireless networks and the development of e-health services facilitated the emergence of e-health systems for mobile patients. In order to support e-health systems with very different QoS needs Vergados (2007) proposed an adaptive resource allocation mechanism for e-health systems based on NGNs and with wireless access. This resource allocation model maps the e-health services onto three different categories of services on the basis of the degree of urgency and interactivity: (1) high priority, corresponding to critical situations;

(2) medium priority for real-time, non-critical e-health services; and (3) low priority, corresponding to non-real-time services. The simulation results showed the effectiveness of implementing these three priority levels associated with different types of flows based on their needs.

In order to transport the different types of flows over NGNs, the Third-Generation Partnership Project (3GPP) defined the Evolved Packet System (EPS) (3GPP 2009; Olsson *et al.* 2009), which is an evolution of the UMTS in order to support an increase in and diversification of traffic while allowing higher flow rates and shorter delays. EPS made it possible to control QoS in the network because of a differentiation of services (Ekström 2009). As with the DiffServ protocol (Babiarz *et al.* 2006), the routing of packets in a flow/session/service is based on a QoS Class Identifier (QCI). Each of the nine QCI categories is characterized by particular delay and packet loss rate values that must be respected by the network using the mechanisms put in place (for example, admission control, priority order, wait queue management, etc.). In this context, the research carried out in Skorin-Kapov and Matijasevic (2010) aimed to define the relevant mapping between e-health services and the QCI categories. This mapping, which is based on the QoS needs of e-health services, allows network operators to efficiently prioritize the flows transported and to take optimal decisions about the allocation of the required network resources. They cited the example of a teleconsultation service between a patient and a doctor, which illustrated how useful this proposition was and demonstrated the operation of QoS (EPS) procedures in the presence of such a mapping.

In the context of NGNs, another QoS management method that may be essential is QoS signaling. Indeed, in NGNs, transporting a service may involve several intermediate networks in addition to access networks. In order to guarantee the end-to-end QoS required by a service (telediagnosis, for instance), each network provider whose network is involved in the transport has to guarantee a certain QoS for the transiting flow. Thus, different network providers must communicate with each other to establish an agreement on the end-to-end QoS that must be guaranteed for the transported flows. This communication may be carried out using signaling protocols, or, more precisely Service Level Negotiation Protocol (SLNP) (Chalouf and Krief 2013). The SLNP defines a variety of technical parameters that may be negotiated between these different network providers for a given service. These parameters, grouped together into a Service Level Specification (SLS), include QoS parameters such as delay, jitter, data rate and packet loss. Further, this protocol may easily be used for the QoS signaling required by e-health services for two main reasons. First, it makes it possible to negotiate different types of guarantees such as instantaneous QoS, scheduled QoS or periodic QoS. This corresponds perfectly to certain e-health services such as telemonitoring. Second, the SLNP could be extended in order to integrate new QoS parameters or new SLS formats such as predefined SLS. Since it uses web services, the format for the SLS negotiated with SLNP can easily be extended through a simple change of the eXtensible Markup Language (XML) schema that describes its format (Mbarek *et al.* 2011).

4.3.2.3. *e-health systems integrating new technologies*

The emergence of new paradigms, such as the Internet of Things (IoT) was accompanied by the development of new technologies that were perfectly adapted to the needs of new services and applications. Among these new technologies, we can cite the examples of Low Power Wide Area Network (LPWAN) and LoRaWAN, in particular, supported by the LoRa Alliance (2018) or even, the 802.11.ah promised by the Wi-Fi Alliance (2018), which were defined for IoT applications with low data rates. 3GPPP also positioned itself in this market with the arrival of the LTE-M and NB-IoT standards (*3GPP* 2018). These new technologies, characterized by a wide range and low energy-consumption, made it possible to access the IP-based network. These technologies could represent an alternative for the transport of some e-health services that do not have very high QoS demands, such as the periodic transmission of certain vital signs in a normal (non-critical) situation.

Other new technologies have emerged over the past few years, such as cognitive radio (CR) or smart radio. Cognitive Radio (Mitola 2006) allows the opportunistic use of the unused part of licensed frequencies by users (devices) who have no priority on these frequencies (no right). These users must, therefore, avoid interfering with primary users (who hold licenses for these frequencies) and free up the frequencies they are using opportunistically as soon as a licensed user states the need to communicate. There are several advantages to using CR that may resolve many problems related to e-health, such as telemonitoring in rural zones (Ouattara *et al.* 2014b). This smart feature allows CR to optimally manage the frequencies used in wireless communication and, thus, allows it to be better performing (flow rate). This makes it possible for this technology, among others, to contribute to the improvement of the quality of the transmissions in e-health systems and to guarantee better performances for e-health services. In this context, Ouattara *et al.* (2017) propose an e-health system based on CR. In this system, the patient's principal device allows the transmission of medical flows toward the IP network (remote servers or contacts' devices) using the CR capacities of this equipment. Since a CR device must first detect free channels before transmitting, data transmissions are not always instantaneous. This could pose a serious problem in the context of urgent or real-time e-health services, which could be coupled with a patient's limited mobility. A spectral prediction function was proposed (Ouattara *et al.* 2012) as a solution to this problem. This function makes it possible to enhance modules to detect and share the spectrum, taking into account the requirements of the e-health services. This function is based on machine learning (Cornuéjols and Miclet 2010) associated with the gray model (Sheng *et al.* 2011) and makes it possible to identify free channels in advance in order to anticipate the frequency changes caused by the patient's mobility or the arrival of primary users into the used channels. The e-health system based on CR must handle QoS problems. Indeed, on the one hand we have e-health flows with different QoS requirements, while on the other hand there is available bandwidth that depends on the channel used and its availability rate. In addition, there is the potential drop in performance of the transmission of flows linked to the change of the transmission channel. Ouattara *et al.* (2014a) proposed a mechanism to reserve frequency resources on the fly to resolve this problem. To guarantee the effectiveness of the channel reservation and to ensure that this reservation responds to the flow's QoS requirements, a mechanism to control the transmission parameters, based on Model Predictive Control (MPC), was used by Zhang and Wan (2012).

Furthermore, e-health systems often involve wireless medical devices such as medical sensors that must communicate with each other and/or with a gateway, such as a smartphone. These wireless radio communications may interfere with each other or even interfere with critical medical devices such as insulin pumps or pacemakers, provoking serious incidents. In this context, using CR is advantageous as it makes it possible to select the frequency bands that have no impact or minimal impact on equipment in the vicinity (Phunchongharn *et al.* 2010; Ouattara *et al.* 2013). For example, Ouattara *et al.* (2013) proposed using CR for critical components of e-health systems (hospital, medical health centers, patient's home, etc.) while defining mechanisms that would minimize interference in order to ensure the smooth working of all sensitive medical equipment.

Using WBAN (Wireless Body Area Networks) in e-health systems could cause a drop in the QoS for these systems, given the limitations of this technology in terms of computation power, memory and communication abilities. In order to improve the performance of e-health systems, especially telemonitoring systems based on WBAN (Almashaqbeh *et al.* 2014) proposed a new approach, where the e-health system was based on two Clouds (a local Cloud for the patient and an external Cloud) and a number of new techniques. These techniques make it possible to reduce congestion, interference and transmission delays, while supporting user mobility. Data classification and aggregation techniques were used in order to avoid transmitting useless data. Further, a strategy to share frequency bands was implemented in order to avoid interference between equipment and users who must communicate. In this e-health system (Almashaqbeh *et al.* 2014), a delay-sensitive routing technique was used to accelerate the routing of e-health data within the local Cloud. Iftikhar and Iftikhar (2014) proposed a new analytical framework in the context of improving QoS within WBANs. This made it possible to test models of traffic as well as QoS mechanisms (for example, ordering strategies, wait queues) that are more realistic in order to find the best conditions that make it possible to obtain the QoS required for e-health services.

4.4. Security of e-health systems

In section 4.2, we presented the architecture as well as the characteristics of e-health systems. As we observed, these systems may be made up of several pieces of equipment localized in different networks and connected by wired and wireless communication devices. As with any information system, e-health systems may be targeted by cyberattacks. This may have serious consequences on the patient's health.

In this section, we will review the importance of security in e-health systems and then describe the principal mechanisms that enable guaranteeing the security of these systems.

4.4.1. *Threats and attacks specific to e-health systems*

Technological advances have contributed significantly to new applications such as e-health. e-health systems include multiple pieces of equipment, devices and objects capable of communicating and exchanging health data. Security threats and a lack of security mechanisms that are adapted to e-health systems have slowed down the adoption of these systems and their commercialization. Indeed, security and privacy were neglected in the earliest e-health systems and there were many attacks that could have compromised these systems and their functioning. For example, consider a diabetic patient with cardiovascular disease. The e-health system for this patient will monitor blood sugar, oximetry and heartbeat. Among the devices used for this telemonitoring are a blood sugar monitor, an insulin pump, an oximeter and an electrocardiogram. In this case, a malicious attack that aims to disrupt or compromise the good functioning of the e-health system can have a fatal consequence for the patient. For example, a malfunctioning piece of equipment may impersonate a blood sugar sensor and always transmit normal values. This will then prevent the e-health system from reacting following an actual abnormal value (that is not transmitted) and may have serious consequences. This explains why the success, adoption and commercialization of these e-health systems depend on the security of these systems and patients' privacy. Thus, security must be guaranteed for the data collected, exchanged, stored and made accessible within e-health applications.

4.4.1.1. *Threats and attacks specific to e-health systems*

The architecture of e-health systems, discussed in section 4.2.1, is composed of two main networks: the patient's local network and the external network. The local network comprises various heterogeneous equipment (e.g. WBAN) and the external network includes equipment that may be in different locations (e.g. remote servers such as those in the Cloud). These equipment are involved in the processing, storage and transmission of private health data and may be sensitive and critical. The multilocation of the data storage and the exchange of this sensitive data over wireless and mobile networks increase the security risks. There are thus multiple threats and attack risks and these may range from affecting the physical security of a device to the security of storage and data transmission. In this section, we will focus on the security of the data exchanged and stored within an e-health service.

In general, security attacks are classified into two types: active attacks and passive attacks (Chalouf 2010). Passive attacks allow unauthorized access to information without modification or with imperceptible changes. Active attacks allow unauthorized access to information with modification. Another category that may be considered in e-health systems is that of threats related to data storage and access (Ming *et al.* 2010) and threats related to data exchange (Shahnaz *et al.* 2011). Given the specific architecture of e-health systems, another category of attacks was discussed in Shihong *et al.* (2017). This consists of security attacks against the local network, namely the WBAN, and attacks targeting exchanges and interactions between the local network and the external network. As for communications with the exterior, we have conventional attacks such as passive listening, identity theft, replay, Denial of Service (DoS), etc. Otherwise, the attacks that target WBANs are more wide ranging, given the specific features of these networks. These attacks may be classified into three categories (Javadi and Razzaque 2013): (a) attacks breaching information secrecy and authentication, where an attacker eavesdrops, replays or spoofs; (b) attacks affecting packet integrity, where the WBAN network is forced to accept false information and (c) attacks threatening the availability of the network (DoS attacks), where the attacker aims to reduce the capacity of the network or to put it out of service. DoS attacks may be carried out at different levels of the protocol stack (Javadi and Razzaque 2013). An example of this is interference, which affects the physical layer; collision, which affects the link level; black holes, which affect the network level; and, desynchronization, which affects the transport level.

4.4.1.2. *Security needs of e-health systems*

The above-mentioned attacks can cause serious problems to users of e-health systems such as patients and elderly. In order to protect such persons against these attacks, the sensitive data exchanged, processed and stored within e-health systems must be protected. Thus, security and privacy protection needs can be divided into three categories (Javadi and Razzaque 2013): privacy and security of access to data, security of data exchanged over the network and security of data storage. In this section, we describe the main security services required by an e-health system.

4.4.1.2.1. Privacy and security of data access

The first category of services concerns privacy and data access security. This includes confidentiality, access control, responsibility, revocability and non-repudiation.

– confidentiality: this is protecting the transmitted data against leakage and disclosure. An attacker may monitor a communication in an e-health system (for example, communication between sensors and the gateway) and eavesdrop on the transmitted information, which they can then use for illegal purposes. To protect the user's privacy, all communications in an e-health system (within the WBAN, between the WBAN and the gateway and with the exterior) must be protected;

– access control: this prevents unauthorized access to patient data. In an e-health system, the patient's medical data may be consulted by different users and parties such as doctors, nurses, pharmacies and insurance companies. For example, some medical information may be used by insurers to apply high rates (Li *et al.* 2010b). Thus, role-based access control to the WBAN and its data must be implemented;

– responsibility: when a user is identified and held responsible for abusing his/her rights to carry out unauthorized actions on patient data;

– revocability: this protects the network against compromised nodes/users. If a user or a node is compromised, they/it must quickly be deprived of all authorizations they/it may have previously received (Li *et al.* 2010b);

– non-repudiation: this ensures that the sender of a message cannot deny being the origin of this message and that the recipient cannot deny having received the message.

4.4.1.2.2. Security of communications

This second category of services is related to the security of communications. This includes integrity, authentication of origin, freshness and availability of data.

– data integrity: it guarantees the non-alteration of the information received. This allows the different devices in an e-health system to verify that the data are authentic and correct and allows them to efficiently treat patients;

– authentication of the data origin: this makes it possible to be sure that the information is sent by a trusted entity. This prevents an illegitimate entity from impersonating a legitimate entity and reporting false data or giving false instructions in order to cause problems to the patient;

– data freshness: it makes it possible to ensure that the data received are fresh. The system must be able to verify that data frames are received in the right order and are not replayed by an attacker in order to disrupt the functioning of the e-health system;

– data availability: means that the patient's medical data must always be available for authorized persons such as a doctor, even in the case of a DoS type attack.

4.4.1.2.3. Security of storage

The third class of services concerns the security of storage and includes confidentiality, integrity and reliability:

– confidentiality: data related to the patient must always remain confidential, not only during transmission but also during storage;

– integrity: in e-health systems, data integrity must always be verified, not only during transmission but also during storage;

– reliability: this means that data related to the patient must be easy to recover in case of a failure in the e-health system (for example, in the case of a compromised node or malicious modification).

4.4.2. *Security management in e-health systems*

The previous sections introduced some attacks that could disrupt the functioning of e-health systems and damage a patient's health, before discussing the main security services that can protect e-health systems. These security services can be guaranteed using security mechanisms that must be implemented. Table 4.4 provides some examples of security mechanisms that may be implemented to resist some malicious attacks.

Attacks and threats	Security services	Possible solutions
Unauthorized access	Key management and trust management	Random distribution of keys Asymmetric encryption
Divulging a message	Confidentiality and privacy	Data encryption Access control
Modification of a message	Integrity and authentication	Hash Message Authentication Code Digital signature
Denial of service	Availability	Detection of intrusion and redundancy
Captured node	Resistance to a node being compromised	Detection of inconsistency Revocation of a node Tamper-proofing
Intrusion	Intrusion detection	Intrusion detection

Table 4.4. *Security required by WBANs and possible solutions (Ng* et al. *2006)*

Concerning the data exchanged within an e-health service, it is essential to guarantee the confidentiality of communications within the local network (between the WBAN nodes and between the WBAN and the gateway) and between the local network and the exterior. The confidentiality of the exchanged data may be guaranteed using encryption (symmetric, asymmetric, elliptical curves, etc.). Data integrity as well as the authentication of the data origin are two security services that are guaranteed together using an encryption mechanism such as Hash Message Authentication Code (HMAC) based on symmetric encryption, or a digital signature based on asymmetric encryption. The digital signature also guarantees non-repudiation. The freshness of the data and the guarantee of data being processed in the right order makes it possible to protect against attacks such as replays and this may be guaranteed through adding a sequence number. As for access control to data, there are several solutions that can be used. These solutions are based on the definition of access rules that allow a user to access data only after authentication. Evered and Bogeholz (2004) provided an example of an access control system based on the roles played by different actors (healthcare professionals, insurance companies, patient, etc.).

The main security services are related to the patient's medical data exchanged at the level of the local network and also at the level of the external network. For communications between the local network and the exterior or within the exterior network (for example, between healthcare actors and Cloud servers), there are many security protocols to protect communications such as the IP Security protocol (Frankel and Krishnan 2011) or the TLS protocol (Dierks and Rescorla 2008) and DTLS protocol (Rescorla and Modadugu 2012). These protocols make it possible to guarantee the confidentiality, integrity, authentication of the data origin, non-repudiation and protection against replay for different types of flows.

However, the processing, storage and exchange of medical data over the local network (essentially between WBAN nodes) require not only mechanisms adapted to guarantee security services but also other security mechanisms in order to protect this sensitive part of e-health systems against specific attacks. For example, Meingast *et al.* (2006) defined an access control system to access medical data in the local network. This system introduced rules that made it possible to respect the patient's privacy. The system defines who can decide what sensors must collect data, the degree of patient involvement (complete or partial) in controlling the data transmitted to the servers as well as the necessary directives. As for the responsibility of the users of medical data, Yu *et al.* (2009) propose a solution that makes it possible to manage this problem. Furthermore, WBAN networks may require different guarantees in terms of data freshness: low freshness, which ensures that packets are processed in the right order, and strong freshness, which guarantees low delays. For example, strong freshness is required for the WBAN synchronization phase (Saleem *et al.* 2009). As seen in Table 4.4, one of the solutions proposed to solve the problem of availability in a WBAN is redundancy. This makes it possible to transfer the functionalities of a deactivated node to an available node. Redundancy is essential when there are nodes carrying out vital operations, but requires solutions to manage secrets when a new node replaces a deactivated or a compromised node. This management aims to authorize a new node to only read future messages exchanged, while preventing the compromised node from doing so (Javadi and Razzaque 2013).

As discussed above, security is very important not only for data transmission but also for data storage. Patient data are often stored on servers that are not monitored by either the patient or the provider of the e-health service, as in the case of Cloud servers. This makes it necessary to encrypt data before storing them on these dedicated servers. In Ermakova and Benjamin (2013), in addition to the encryption of data stored on Cloud servers, the authors also introduce two new functionalities that aim to strengthen the confidentiality of stored medical data as well as the patient's privacy. These two functionalities may be based on a shared secret and are:

(1) "unlinkability" between the medical data and the patient; (2) confidentiality of the patient's identity. Other new functionalities with the same goal, such as anonymization, were also defined (Li *et al.* 2010a).

A large part of the security services required by e-health systems, such as confidentiality, integrity and authentication, may be carried out using cryptographic mechanisms such as symmetric and asymmetric encryptions. We recall that symmetric encryption is based on a single secret key, which can only be known by both ends of the communication. This encryption has the advantage of being fast and not being resource intensive (processor time, memory and energy), but it requires an efficient and secure key distribution system. Asymmetric encryption is based on a pair of keys: the public key, distributed to everyone and the private key, which is kept secret. The advantages of this type of encryption are that it is more robust compared to symmetric encryption and the distribution of keys is quite simple. In IP networks, the equipment and machines have sufficient processing capacity, memory space and energy sources, which favors the use of asymmetric encryption techniques. Nevertheless, this type of encryption (asymmetric) is not well adapted for networks composed of nodes with limited capabilities (CPU, memory, energy, etc.) such as WBANs (Carman *et al.* 2000; Vijayalakshmi 2011). It is, therefore, more advantageous to use symmetric encryption to secure such a network and to use new techniques for distributing symmetric keys between the nodes of these networks. Several research projects have examined how to define methods for generating and exchanging symmetric keys that are well adapted to WBANs, such as methods based on physiological data such as an ECG (Mana *et al.* 2009; Huawei *et al.* 2015; Sammoud *et al.* 2018).

4.5. Conclusion

E-health is an emerging technological environment. To be accepted by healthcare professionals and, more importantly, by patients, the problems of the e-health systems must be resolved. The most important of these problems are quality and reliability as well as security and privacy. This chapter first described e-health systems and their functioning and then summarized the major efforts in this field. The projects described in this chapter had the following objectives: (1) improving the QoS in e-health systems in order to satisfy the different needs of various e-health services; and (2) guaranteeing essential security services to protect these systems against possible attacks.

While a large amount of research has been carried out on guaranteeing good QoS and security for e-health systems, it is still possible to add to this. Regarding the QoS, it is not only possible to refine some of the propositions described in this chapter but also to propose a global framework to orchestrate the mechanisms used in the two main networks (local and external) in order to have good end-to-end QoS. As for security, there remains a lot to be done in order to improve the robustness of the cryptographic mechanisms used. Blockchain, for example, represents a good alternative that would make it possible, among other things, to store and share medical data in a secure manner between different actors in an e-health system. We have also to note that the use of cryptographic mechanisms requires exchanges and computations that may result in significant energy consumption and additional delays. This may have a negative impact on the quality of the e-health service (for example, transmission delays, autonomy of a mobile e-health system, etc.). There are, therefore, compromises that must sometimes be made between these two important aspects: QoS and Security.

4.6. References

3GPP (2018). Narrow band IoT [Online]. Available at: https://www.3gpp. org/news-events/3gpp-news/1785-nb_iot_complete [Accessed 10 March 2020].

3GPP TS 22.278 (2009). Service requirements for the evolved packet system, V9.5.0, Release 9.

Agence française de la santé numérique (2010). Le décret du 19 octobre 2010 relatif à la Télémédecine [Online]. Available at: https://www.legifrance.gouv.fr/affichTexte.do?cidTexte=JORFTEXT000022932449&categorieLien=id, [Accessed 10 March 2020].

Almashaqbeh, G., Hayajneh, T., Vasilakos, A.V., and Mohd, B.J. (2014). QoS-aware health monitoring system using cloud-based WBANs. *Journal of Medical Systems*, 38(121).

Babiarz, J., Chan, K., and Baker, F. (2006). Configuration guidelines for DiffServ service classes [Online]. Available at: https://tools.ietf.org/html/rfc4594, [Accessed 10 March 2020].

Carman, D.W., Kruus, P.S., and Matt, B.J. (2000). Constraints and approaches for distributed sensor network security. Report, NAI LABS.

Chalouf, M.A. (2010). Communications security in embedded systems. In *Communicating Embedded Systems*, Krief, F. (ed.). ISTE Ltd, London and Wiley, New York.

Chalouf, M.A. and Krief, F. (2013). A secured, automated and dynamic service level negotiation. *Concurrency and Computation: Practice and Experience*, 25(2), 180–202.

Chronious project (2012). *European Commission's 7th Framework Programme (Personal Health Systems)* [Online]. Available at: http://www.chronious.eu/ [Accessed 10 March 2020].

Chu, Y. and Ganz, A. (2004). A mobile teletrauma system using 3G networks. *IEEE Transactions on Information Technology in Biomedicine*, 8(4), 456–462.

Chu, Y. and Ganz, A. (2007). WISTA: A wireless telemedicine system for disaster patient care. *Mobile Networks and Applications*, 12, 201–214.

Cornuéjols, A. and Miclet, L. (2010). *Apprentissage Artificiel. Concepts et algorithmes*. Eyrolles, Paris.

Dierks, T. and Rescorla, E. (2008). The Transport Layer Security (TLS) protocol. Version 1.2. Memo, IETF, 5246.

Ekström, H. (2009). QoS control in the 3GPP evolved packet system. *IEEE Communications Magazine*, 47(2), 76–83.

Ermakova, T. and Benjamin, F. (2013). Secret Sharing for Health Data in Multi-provider. *IEEE 15th Conference on Clouds Business Informatics*. Vienna.

Evered, M. and Bogeholz, S. (2004). A case study in access control requirements for a health information system. *The Second Workshop on Australasian Information Security, Data Mining and Web Intelligence, and Software Internationalisation*. Dunedin.

Frankel, S. and Krishnan, S. (2011). IP Security (IPsec) and Internet Key Exchange (IKE) document roadmap. Memo, IETF, 6071.

Gállego, J.R., Hernández-Solana, A., Canales, M., Lafuente, J., Valdovinos, A., and Fernández-Navajas, J. (2005). Performance analysis of multiplexed medical data transmission for mobile emergency care over the UMTS channel. *IEEE Transactions on Information Technology in Biomedicine*, 9(1), 13–22.

Garawi, S., Istepanian, R.S.H., and Abu-Rgheff, M.A. (2006). 3G wireless communi-cations for mobile robotic tele-ultrasonography systems. *IEEE Communications Magazine*, 44(4), 91–96.

Hamdi, O., Chalouf, M.A., Ouattara, D., and Krief, F. (2014). eHealth: Survey on research projects, comparative study of telemonitoring architectures and main issues. *Journal of Network and Computer Applications*, 46, 100-112.

HealthService 24 (eTEN-517352) (2006). [Online]. Available at: http:// www. healthservice24.com/ [Accessed 10 December 2012].

Huawei, Z., Chi, C., Jiankun, H., and Jing, Q. (2015). Securing Body Sensor Networks with Biometric Methods: A New Key Negotiation Method and a Key Sampling Method for Linear Interpolation Encryption. *International Journal of Distributed Sensor Networks,* 11(8), 11.

Iftikhar, M. and Iftikhar, A. (2014). A novel analytical model for provisioning QoS in body area sensor networks. *International Workshop on Body Area Sensor Networks*. BASNet, Halifax.

Istepanian, R.S.H., Alinejad, A., and Philip, N. (2012). Medical quality of service (m-QoS) and quality of experience (m-QoE) for 4G-health systems. In *Multimedia Networking and Coding.* Farrugia, R.A. and Debono, C.J. (eds). IGI Global, Hershry, 359–376.

Javadi, S.S. and Razzaque, M.A. (2013). Security and privacy in wireless body area networks for health care applications. In *Wireless Networks and Security*, Khan, S. and Pathan, A.S.K. (eds). Springer-Verlag, Berlin, 165–187.

Konstantas, D., Van Halteren, A., Bults, R., Wac, K., Widya, I., Dokovsky, N., Koprinkov, G., Jones, V., and Herzog, R. (2004). Mobile patient monitoring: The MobiHealth system. *Studies in Health Technology and Informatics*, 103(5), 307–314.

Kumar, S., Kambhatla, K., Lifson, M., and Xaio, Y. (2008). Ubiquitous computing for remote cardiac patient monitoring: A survey. *International Journal of Teleme-dicine and Applications*, 2008. 459185. 10.1155/2008/459185.

Li, M., Yu, S., Ren, K., and Lou, W. (2010a). Securing personal health records in cloud computing: Patient-centric and fine-grained data access control in multi-owner settings. In *Security and Privacy in Communication Networks.* Springer, Berlin, 89–106.

Li, M., Lou, W., and Ren, K. (2010b). Data security and privacy in wireless body area networks. *Wireless Communications*, 51–58.

LoRa Alliance (2018). Wide area networks for IoT [Online]. Available at: https://lora-alliance.org/about-lorawan [Accessed 10 March 2020].

Mana, M., Feham, M., and Bensaber, B.M. (2009). SEKEBAN (Secure and Efficient Key Exchange for wireless Body Area Network). *International Journal of Advanced Science and Technology*, 12, 45–60.

Martinez, I., Garcia, J., and Viruete, E.A. (2008). Resources variability in M-health services: An adaptive method for QoS control. *5th IEEE Consumer Communications and Networking Conference*. IEEE, Las Vegas, 829–833.

Mbarek, N., Chalouf, M.A., and Krief, F. (2011). Towards global service level guarantee within autonomic computing systems. In *The 12th IFIP/IEEE International Symposium on Integrated Network Management*, IFIP/IEEE. Dublin.

Meingast, M., Roosta, T., and Sastry, S. (2006). Security and privacy issues with health care information technology. *The 28th IEEE EMBS Annual International Conference*. New York.

Ming, L., Wenjing, L., and Kui, R. (2010). Data security and privacy in wireless body area networks. *IEEE Wireless Communications*, 7(1), 51–58.

Mitola, J. (2006). *Cognitive Radio Architecture: The Engineering Foundations of Radio*. John Wiley & Sons, New York.

Moulin, T. and Simon, P. (2016). E-santé. Objets connectés et télémédecine. *Correspondances en Onco-Hématologie*, 11(2), 115–120.

Navarro, E.A.V., Mas, J.R., Navajas, J.F., and Alcega, C.P. (2006). Performance of a 3G-based mobile telemedicine system. *The 3rd IEEE Consumer Communications and Networking Conference*. IEEE, Las Vegas, 1023–1027.

Navarro, A., Ceccaroni, L., Velickovski, F., Torrellas, S., Miralles, F., Allison, B.Z., and Faller, J. (2011). Context-awareness as an enhancement of brain-computer interfaces. *International Workshop on Ambient Assisted Living*. Torremolinos.

Ng, H.S., Sim, M.L., and Tan, C.M. (2006). Security issues of wireless sensor networks in healthcare applications. *BT Technology Journal*, 24, 138–144.

Ouattara, D., Chalouf, M.A., Hamdi, O., and Krief, F. (2014a). A QoS-control framework for medical multimedia data transmission in CRN environment. *IEEE Symposium on Computers and Communication*. IEEE, Madeira, 1–7.

Ouattara, D., Chalouf, M.A., Krief, F., and Hamdi, O. (2014b). Wireless broadband opportunities through TVWS for networking in rural areas of Africa. *EAI Endorsed Transactions on e-Business*.

Ouattara, D., Chalouf, M.A., and Krief, F. (2017). Towards cognitive radio based e-health systems. *International Journal of Autonomous and Adaptive Communications Systems*, 12(1), 1–33.

Ouattara, D., Krief, F., Chalouf, M.A., and Hamdi, O. (2012). Spectrum sensing improvement in cognitive radio networks for real-time patients monitoring. *The 3rd International Conference on Wireless Mobile Communication and Healthcare*. Paris.

Ouattara, D., Quach, T., Krief, F., and Chalouf, M.A. (2013). Mitigating the hospital area communication's interference using cognitive radio networks. *15th IEEE International Conference on e-Health Networking, Application & Services*. IEEE, Lisbon.

Perakis, K. and Koutsouris, D. (2009). Third generation (3G) cellular networks in telemedicine: technological overview, applications, and limitations. In *Handbook of Research on Distributed Medical Informatics and E-Health*, Lazakidou, A.A., Siassiakos, K.M. (eds). IGI Global, New York, 241–259.

Phunchongharn, P., Hossain, E., Niyato, D., and Camorlinga, S. (2010). A cognitive radio system for e-health applications in a hospital environment. *IEEE Wireless Communications*, 17(1), 20–28.

Qiao, L. and Koutsakis, P. (2008). Guaranteed bandwidth allocation and QoS support for mobile telemedicine traffic. *IEEE Sarnoff Symposium*. IEEE, Princeton, 1–5.

REACTION Project (2014). *European Commission's 7th Framework Programme (Personal Health Systems), remote accessibility to diabetes management and therapy in operational healthcare networks* [Online]. Available at: http://www.reaction-project.eu/ [Accessed March 2018].

Rescorla, E. and Modadugu, N. (2012). Datagram transport layer security version 1.2. Mémo, IETF, 6347.

Rocha, A., Martins, A., Freire Junior, J.C., Kamel Boulos, M.N., Vicente, M.E., Feld, R., Van de Ven, P., Nelson, J., Bourke, A., ÓLaighin, G., Sdogati, C., Jobes, A., Narvaiza, L., and Rodríguez-Molinero, A. (2011). Innovations in health care services: The CAALYX system. *International Journal of Medical Informatics*, 82(11), 307–320.

Saleem, S., Ullah, S., and Yoo, H.S. (2009). On the security issues in wireless body area networks. *International Journal of Digital Content Technology and its Applications*, 3(3), 178–184.

Salehi, L. and Mirhadi, P. (2008). E-health applications implementation considerations. *4th Kuala Lumpur International Conference on Biomedical Engineering*. Kuala Lumpur, 21, 858–861.

Sammoud, A., Hamdi, O., Chalouf, M.A., and Bouallegue, A. (2018). A new protocol for an efficient and green biometric-based security key establishment in WBANs. *International Wireless Communications and Mobile Computing Conference*. Limassol.

Shahnaz, S., Sana, U., and Kyung, S.K. (2011). A study of IEEE 802.15.4 security framework for wireless body area networks. *Sensors*, 11, 1384–1395.

Sheng, C., Lin, Y., Gengwu, Z., Cheng, Z., Shijun, D., and Chao, D. (2011). Short-term wind power prediction based on combined grey-markov model. *International Conference on Advanced Power System Automation and Protection*. Beijing, 3, 1705–1711.

Shihong, Z., Yanhong, X., Honggang, W., Zhouzhou, L., Shanzhi, C., and Bo, H. (2017). A survey on secure wireless body area networks. *Journal of Security and Communication Networks*, 2017, 9.

Simon, P. (2010). New organizations and healthcare management with telemedicine. *Médecine des maladies métaboliques*, 4(3), 251–256.

Skorin-Kapov, L. and Matijasevic, M. (2010). Analysis of QoS requirements for e-health services and mapping to evolved packet system QoS classes. *International Journal of Telemedecine Applications*, 2010, 18.

Skorin-Kapov, L., Mosmondor, M., Dobrijevic, O., and Matijasevic, M. (2007). Application-level QoS negotiation and signaling for advanced multimedia services in the IMS. *IEEE Communications Magazine*, 45(5), 108–116.

SOPRANO Project (2011). Service-oriented Programmable Smart Environments for Older Europeans [Online]. Available at: http://imu.ntua.gr/project/soprano-service-oriented-programmable-smart-environments-older-europeans [Accessed 10 March 2020].

Vergados, D.D. (2007). Simulation and modeling bandwidth control in wireless healthcare information systems. *International Transactions of The Society for Modeling and Simulation*, 83(4), 347–364.

Vijayalakshmi, B. (2011). A zero-knowledge authentication for wireless sensor networks based on congruence. *The Third IEEE International Conference on Advanced Computing*. Chennai.

Vouyioukas, D., Maglogiannis, I., and Komnakos, D. (2007). Emergency M-health services through high-speed 3G systems: simulation and performance evaluation. *International Transactions of the Society for Modeling and Simulation*, 83(4), 329–345.

Wi-Fi Alliance (2018). IEEE 802.11ah standard [Online]. Available at: https://www.wi-fi.org/discover-wi-fi/wi-fi-halow.

Yu, S., Ren, K., Lou, W., and Li, J. (2009). Defending against key abuse attacks in KP-ABE enabled broadcast systems. *5th International ICST Conference on the Security and Privacy in Communication Networks*. Athens.

Zhang, J. and Wan, S. (2012). A review of explicit model predictive control. *31st Chinese Control Conference*. Hefei.

Zvikhachevskaya, A., Markarian, G., and Mihaylova, L. (2009). Quality of service consideration for the wireless telemedicine and e-health services. *IEEE Wireless Communications and Networking Conference*. Budapest, 1–6.

5

Quality of Service Management in Wireless Mesh Networks

Hajer BARGAOUI, Nader MBAREK and Olivier TOGNI

LIB, University of Bourgogne Franche-Comté, Dijon, France

5.1. Introduction

Telecommunication networks today make it possible to respond to an essential need for all users: the ability to communicate in one's daily and professional life. Technologies are constantly evolving at an impressive pace. Following the invention of landline telephones and wired networks in general, the past few years have seen the emergence of various wireless technologies. This evolution has made it possible to bring down the cost of installing networks and of extending connectivity to areas where establishing wired networks was difficult. Wireless networks led to the rise of several categories of services with different characteristics. Centralized wireless networks using access points were among the first kind of wireless networks to be deployed. These were followed by *ad hoc* networks that were brought in as a distributed solution with all nodes playing the same role and offering the functionalities of a relay node. In recent years, however, researchers have largely focused on a new category of networks, called wireless mesh networks (WMNs). These networks represent a new generation of wireless networks that have inherited features from *ad hoc* networks but offer wider coverage, greater flexibility in deployment and better performance as a result of the meshing.

This evolution of telecommunication networks was accompanied by rising demands from users across different services and multimedia applications. Indeed,

Service Level Management in Emerging Environments,
coordinated by Nader MBAREK. © ISTE Ltd 2020.

today we see a large diversity in the applications available to users in existing wireless networks. We can cite here the example of voice over Internet protocol (VoIP) applications, video on demand, online games, video-conferencing, file transfers, etc. These applications have different characteristics and different QoS requirements. A common challenge related to effectively responding to user needs when these applications are deployed is ensuring there is satisfactory and appropriate QoS for these various applications during the routing of the corresponding traffic in a WMN environment.

In section 5.2, we will first study the characteristics of WMNs. Section 5.3 is then dedicated to the definition of quality of service (QoS) in a WMN environment. Section 5.4 defines the concepts of QoS-based routing in this kind of environment. Finally, section 5.5 describes HQMR, a QoS-based routing protocol defined for WMNs.

5.2. WMNs: an overview

5.2.1. *Definition of a WMN*

Ad hoc networks were defined to eliminate the centralization in wireless networks that use access points and to provide a solution that was based on a distributed and flexible architecture. This kind of network has no hierarchy – all nodes in the network have the same functionalities and are able to transmit data toward other nodes, thus playing the role of relay nodes. Communications between two nodes that are not directly in communication range may be ensured through several intermediate nodes. WMNs were developed while respecting these principles and are defined as a non-centralized wireless network with multihop communication, characterized by mesh topology that ensures mutual interconnection across all nodes.

Unlike *ad hoc* networks, WMNs (Akyildiz and Wang 2005; Yan Zhang 2006) may have a fixed backbone formed by nodes that are interconnected through radio links. This backbone ensures not only that client nodes are interconnected but also provides connection to external networks such as the Internet through gateway nodes.

5.2.2. *Architecture of a radio mesh wireless network*

A WMN mainly contains two types of nodes: mesh routers and mesh clients. The mesh routers are conventional routers with additional meshing functionalities. All these nodes together form the backbone of the WMN. These mesh routers are static nodes that have no energy constraints.

Mesh clients represent the users of the WMNs and they may be fixed nodes or mobile nodes with much simpler meshing functionalities than those of a mesh router. The architecture of a WMN can be categorized into three different models based on the functionalities of nodes: client architecture, infrastructure architecture (backbone), and hybrid architecture, as shown in Figure 5.1.

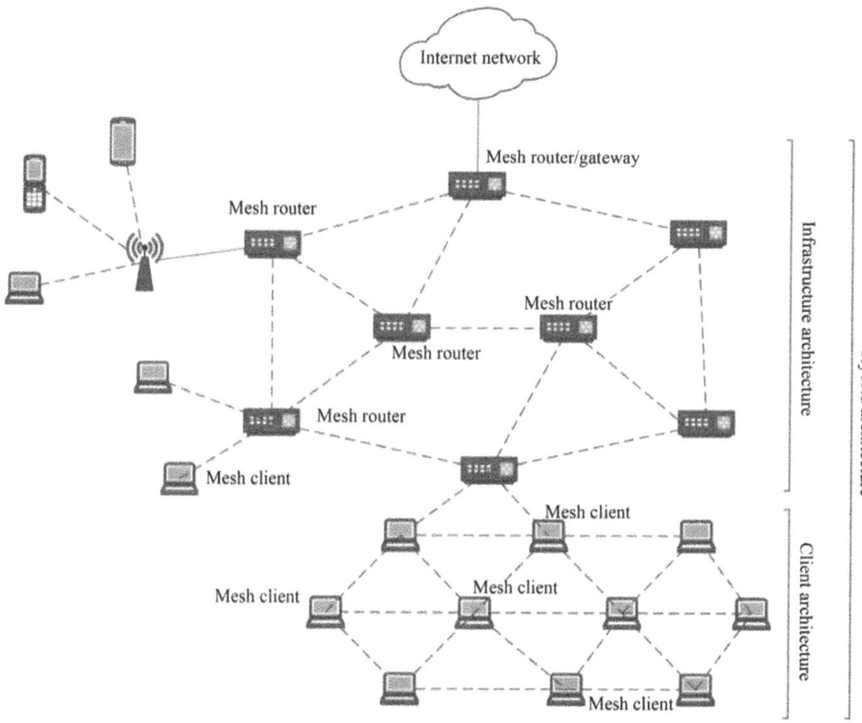

Figure 5.1. *The different architectures of a wireless mesh network*

Client architecture is uniquely formed by a set of client mesh nodes. This enables point-to-point communication and each node must carry out the role of a client as well as that of an intermediate node to route data toward other users. This kind of architecture is similar to that of an *ad hoc* network. The second type of WMN architecture, infrastructure architecture (backbone), has two kinds of mesh nodes, namely mesh routers and clients. The mesh routers form an auto-configurable infrastructure, which can provide different services necessary to mesh clients. In addition, some mesh routers can carry out additional gateway functionalities that enable interconnection with infrastructure in other, external, networks such as the Internet.

The third type of architecture, hybrid architecture, is defined as being a combination of client architecture and infrastructure architecture. In hybrid architecture, the client nodes can communicate with each other either directly, passing through intermediate client nodes, or through the backbone of the WMN, by connecting to a mesh router. In addition, clients have access to external networks because of the presence of infrastructure, while also having greater mobility. Given its flexibility, hybrid architecture is the focus of many research projects (Siddiqui *et al.* 2008; Pirzada *et al.* 2009; Kiani *et al.* 2015; Rao *et al.* 2015).

5.2.3. *Characteristics of a WMN environment*

Over the past few years, WMNs have aroused the interest of many researchers because of the various advantages they present (Akyildiz *et al.* 2005; Nunez-Martinez and Mangues-Bafalluy 2010; Naveen and Vasanth 2017). They are characterized by being easy to deploy, allowing network connectivity to be extended at low costs to zones that are difficult to access. Extending the coverage of a conventional wireless network has proven to be a complex and costly task since adding an access point requires connection to some existing infrastructure. However, this problem can be overcome by WMNs because of their capacity for multihopping communication. In this case, a network can simply be extended through the addition of new nodes in order to ensure a relay effect. This enhances the scalability of the network. Furthermore, a WMN is a self-configuring and self-organizing network, just like *ad hoc* networks. A mesh node only requires a very simple initial configuration before being used in the network. Consequently, the network is capable of configuring itself and organizing itself whenever there is any modification in its topology, without requiring any human intervention. Following a failure or malfunction in a node, the network is able to continue working without any interruption, relaying information across other intermediate mesh nodes and, thus, improving the network's availability. Further, a WMN is characterized by its interoperability with other existing networks. The additional gateway functionalities implemented in a few mesh routers makes it possible to access other types of networks such as WiMAX (Worldwide Interoperability for Microwave Access) networks, ZigBee networks or cellular networks.

5.2.4. *Standards for WMNs*

5.2.4.1. *IEEE 802.11s technology*

The IEEE 802.11s standard (IEEE 2011) was developed to offer mesh functionalities to IEEE 802.11 networks. This was an extension of the 802.11 network that offered a wireless distribution system. The different access points are interconnected through a WMN for multihopping communication. This defines a layer 2 routing protocol that discovers the routes required for routing data between the different nodes in the network, namely the HWMP (hybrid wireless mesh protocol). This protocol adopts a radio metric (see section 5.4.2.4) corresponding to the quantity of radio resources consumed during the transmission of one frame over a link. The HWMP is a hybrid protocol based on a reactive subprotocol, i.e. the RM-AODV (radio-metric *ad hoc* on-demand distance vector) protocol for communications within the infrastructure of the WMN and a proactive tree-based subprotocol to route data to and/or from external networks through the mesh gateway (MG).

The RM-AODV protocol is an extension of the AODV routing protocol that takes into consideration the use of MAC addresses instead of IP addresses and also takes into account a radio metric in the route discovery mechanism. This reactive protocol may be simultaneously run with the proactive HWMP routing protocol when a node (usually the MG) is configured to be the root of the routing tree formed by the different mesh nodes. The proactive mode can use two different mechanisms. The first is based on the periodic broadcast of PREQ (path request) messages by the root. When an intermediate mesh node receives this message, it creates or updates the route to the root. The root then sends out a PREP (path reply) message for each PREQ received, thus establishing a route from the root to each transmitting mesh node. In the second mechanism, the root node periodically broadcasts a RANN (root announcement) message in the network. The information contained in a RANN message is used to broadcast the value of the route metric to the root. When this message is received, each mesh node creates or updates the route to the root. It then sends a unicast PREQ message toward the root via the mesh node from which it received the RANN message. The root sends a PREP message in response to each PREQ message received. Thus, the PREQ message ensures a route is created from the root to the mesh node and the PREP message creates the route from the mesh node to the root. Once the routing tree is constructed, each source mesh node sends its data to the root, which relays the message to the destination mesh node or the external network.

The 802.11s standard (IEEE 2011) was published in 2011 to define the characteristics of the physical layer and the MAC sublayer for a multihop mesh architecture. Subsequently, in 2012, this standard was incorporated into the revised version of the IEEE 802.11 standard (IEEE 2012a).

5.2.4.2. *IEEE 802.16j technology*

IEEE 802.16 technology was developed mainly for metropolitan networks. This technology makes it possible to provide a broadband Internet connection over a coverage area up to several kilometers. It supports point-to-multipoint communication, where mobile stations must connect directly to the base station. A new working group was formed in 2005 to overcome this centralization. The work carried out within this group gave rise to the new standard IEEE 802.16j (IEEE 2009), which was published in 2009. This standard defines a new mode, the multirelay mode. In a multirelay IEEE 802.16 network, traffic between a base station and a mobile station may be relayed by nodes called relay stations. This mode is characterized by a tree-based topology and by connections based on multihop relays between the different stations of the network and the base station.

In 2012, this standard was incorporated into the revised version of the IEEE 802.16 standard (IEEE 2012b).

5.2.5. *Domains of applications*

As a result of their characteristics, WMNs were adopted to meet the requirements of various types of applications (Akyildiz *et al.* 2005). These networks were considered to be an effective substitute for wired networks as well as wireless networks that used access points. WMNs make it possible to provide an Internet connection in different rural or urban areas that are difficult to access. This type of network, therefore, makes it possible to provide better coverage because of its ease of deployment and flexibility, while also overcoming the problem of the high cost of cables, especially those of optical links (Aruba 2018a; Open-Mesh 2018; Open Garden 2018).

A second area of application for WMNs is illustrated through the supply of network connectivity to corporate premises to reduce installation costs (Ruckus Wireless 2013; Blackbox 2018a). This network connectivity may also extend to several buildings in case wiring is prohibited or difficult. In addition, the use of a WMN in corporate network infrastructure provides better robustness against failures and offers considerable flexibility. The network can be easily developed when the size of the company increases because of the scalability of the deployment of mesh networks.

The field of application of WMNs also extends to rescue and emergency services (Portmann and Pirzada 2008; Rodas-Andrade *et al.* 2017; Firetide 2018a; Strix 2018a). In the event of an incident or disaster, the efficiency of the rescue operations often depends on the quality of the communications that can be established between the different units. WMNs offer quick deployment of a communication network in order to ensure data sharing between teams or persons who may need to move around. This deployment can be carried out without any prior infrastructure. We can cite here the example of the Firechat app (Open Garden 2018), which was used to ensure communication between people during some natural disasters, in particular during the Kashmir floods in 2015. This application uses a mesh network set up between users' mobiles instead of using the Internet. Each mobile that uses Firechat relays the information to other mobiles equipped with Firechat.

We can also cite the example of the "broadband home networking" application. This app uses a set of equipment, such as a projector, printer, television or computer, which can communicate with each other through a network within a house. This application was initially implemented using IEEE 802.11 Wi-Fi technology. However, installing access points can be a costly undertaking and does not cover the entirety of the zone dedicated to this type of application. In addition, two access points can only communicate with each other through a wired local network. It was possible to solve all these problems through the use of a WMN that offered a much cheaper and simpler deployment. In this sense, we can also cite the example of Google Wi-Fi technology (Google Store 2018), which allows wide coverage thanks to a mesh router system.

In addition, WMNs can also be used in several other areas such as transportation to provide Internet access to travelers and to install surveillance cameras (Firetide 2018b; Strix 2018b) in building automation and in the medical sector (Prabakar *et al.* 2012; Aruba 2018b) to facilitate the flow of surveillance and diagnostic data between patients and health professionals. WMNs have also been used in recent years as communication infrastructure in the context of Smart cities and the development of the Internet of Things (IoT) (Liu *et al.* 2017; Malandra and Sansò 2017; Purri *et al.* 2017). CleanWiFi (Andrés 2016) is a technology that has been developed to control and share data on air quality and other environmental parameters with residents of a Smart city in order to raise awareness of the consequences of a degradation of these parameters. CleanWifi is based on a mesh Wi-Fi network to interconnect users with the environmental monitoring system. A system based on an IoT environment (Plageras *et al.* 2016) was also proposed to provide omnipresent medical surveillance. This is based on a WMN to interconnect the different components required (e.g. sensors, cameras, etc.) for surveillance and the rapid transfer of medical information about the patient, whatever their location.

5.3. QoS in WMNs

QoS in a WMN is an important challenge in effectively meeting the different requirements of various existing applications in our networks such as real-time interactive applications (VoIP, video conferencing, etc.), streaming applications (video on demand (VoD), Internet Protocol television (IPTV), etc.), data transfer, etc. A primary challenge in WMNs is, therefore, the need for good resource management while also guaranteeing a satisfactory QoS level for all types of traffic. As a result, several research studies have been turned in this direction to propose solutions that are adapted to this type of environment.

5.3.1. *QoS in networks*

In telecommunications networks, QoS aims to achieve better communication behavior to ensure the proper routing of its content while also adopting a more efficient use of network resources. According to the standardization body, ITU-T, QoS is defined as "the overall effect produced by the quality of operation of a service, which determines the degree of satisfaction of the user of the service" (International Telecommunication Unions 2007). In RFC 2386 (Crawley *et al.* 2008), QoS was characterized as a set of requirements that were to be met by the network to transport traffic from a source to a destination. These requirements can be translated into a set of prespecified attributes, measurable in terms of end-to-end delay, jitter representing the delay variation, bandwidth and packet loss ratio. It must also be noted that QoS requirements differ based on application type. For real-time applications, such as VoIP, the most important attributes are the end-to-end delay for a packet as well as the delay variation (jitter). With regard to elastic traffic such as file transfer or messaging, the reliability of the communication is most important and, thus, requires a very low packet loss ratio.

5.3.2. *QoS constraints in WMNs*

Due to some properties inherent to WMNs, they are subject to certain limitations with respect to guaranteeing satisfactory QoS for different applications. A WMN is characterized by the unreliability of its radio links, where the signal may suffer considerable attenuation because of the distance it must travel, and the presence of various obstacles and interferences that may occur. These interferences are a significant restriction on wireless communications, sometimes resulting in a considerable drop in quality or even in the transmissions being cut off completely. This radio characteristic of links also involves a limited bandwidth in a WMN, given that each radio link is shared between different neighboring nodes. Thus, the absence of a centralized monitoring of the quality of the links and the resources available in a WMN requires the definition of new QoS mechanisms that are adapted to the distributed nature of these networks in order to respond to the requirements of applications in a WMN environment.

Another QoS challenge in a WMN arises from the heterogeneity of this type of network. Mesh routers differ from mesh clients in the degree of mobility and the constraints around energy resources. In addition, the dynamism of the mesh topology following the addition or removal of a node from the network can break the route that was pre-established by the routing protocol used. Thus, given the diversity of the applications used in these networks and the restrictions present in a WMN, a routing solution that is more appropriate for this type of network is seen to be necessary to respond to the QoS requirements of each application.

5.3.3. *QoS mechanisms in WMNs*

Several mechanisms have been proposed to meet QoS requirements in a WMN, both at the network layer and at the data link layer.

5.3.3.1. *QoS in the IEEE 802.11s standard*

In order to better manage the resources available in a multihop radio network while also taking into account the constraints of multimedia applications, the IEEE 802.11s standard proposed mechanisms for controlling access to the medium. The default access mechanism, enhanced distributed channel access (EDCA), and the optional access mechanism, mesh coordination function coordinated channel access (MCCA), were defined in the standard for this purpose. Other optional optimizations have also been defined, such as the intra-mesh congestion control.

The EDCA technique (IEEE 2011) is the default mechanism to access the channel for the IEEE 802.11s standard. This technique differentiates services in the MAC sublayer by defining four categories of access classified in decreasing order of priority: the "Voice" class (AC_VO), the "Video" class (AC_VI), the "Best-effort" class (AC_BE) and the "Background" class (AC_BK). Each of these four classes is assigned a wait queue at every node. These wait queues are characterized by different priorities because of the specification of appropriate values for the transmission parameters. The higher the priority for that class, the shorter the wait time.

The MCCA technique (IEE 11) is an optional access method for the IEEE 802.11s standard. This allows the node to access the channel at given times to avoid contention. This technique is based on the concept of period reservation, where a given node has the highest priority for accessing the channel. Consequently, all neighboring nodes will abstain from accessing the channel as long as the priority node is present. As concerns the mechanism to control intra-mesh congestion, this is an optional mechanism implemented for each node to actively monitor the local use of the channel. If any congestion is detected at a node following an overflow at the level of the wait queue, then this node informs its neighborhood so that each of them can then adjust their flow accordingly.

5.3.3.2. *QoS in the IEEE 802.16j standard*

To guarantee QoS, the 2012 revision of the IEEE 802.16 standard offers a scheduling mechanism for the MAC sublayer level. This mechanism is based on the differentiation of services by assigning each type of flow to a specific class.

Consequently, the method used by the base station for allocating the required resources is defined based on the priority of the class. Five different classes are defined in this regard, namely the Unsolicited Grant Service (UGS) class, the real-time Polling Service (rtPS) class, the non-real time Polling Service (nrtPS) class, the extended real time Polling Service class (ertPS) and the Best Effort (BE) service class.

The UGS class represents real-time flow transporting fixed-size data packets at regular intervals, such as E1 (European carrier: 2 Mbit/sec) and VoIP, without deleting silence. This class of service is subject to QoS restrictions in terms of the minimum reserved bandwidth and the maximum tolerated latency and jitter parameters. The rtPS class includes real-time service flows transporting data packets of variable sizes at regular intervals, such as a Moving Picture Experts Group (MPEG) video type flow. This is subject to QoS constraints in terms of the minimum bandwidth reserved and the maximum latency tolerated. The ertPS class combines some characteristics of both the previous classes, namely the UGS class and the rtPS class. It includes real-time service flows transporting data packets of variable size, such as VoIP with deletion of silence. It is also subject to QoS constraints in terms of minimum reserved bandwidth and maximum latency tolerated. The nrtPS class was designed for non-real-time flows, which transport data packets of variable size with a variable interpacket interval, such as the File Transfer Protocol (FTP). QoS constraints for this class are only defined in terms of the minimum reserved bandwidth. The BE class defines the set of BE applications where no QoS guarantee is required, as in electronic messaging, for instance.

5.3.3.3. *Other QoS mechanisms in WMNs*

In parallel with the mechanisms defined by the IEEE 802.11s and IEEE 802.16j standards, there are several other mechanisms that have been proposed to guarantee a satisfactory QoS level in WMNs.

Thus, some mechanisms work in conjunction with a routing protocol and others are implemented independently. The admission control mechanism, for example, helps to avoid the nodes and the network being congested in general, thereby contributing indirectly to guaranteeing a satisfactory QoS level (Shen *et al.* 2008; Dhurandher *et al.* 2015; Khasawneh *et al.* 2016). Shen *et al.* (2008) introduced an algorithm for WMNs called admission control algorithm (ACA), which makes it possible to offer appropriate QoS for real-time flows. ACA is based on the "channel busyness ratio" technique (Zhai *et al.* 2004) to estimate available bandwidth. In order to avoid saturation of the network, ACA also deploys a rate adaptation mechanism for elastic flows. Thus, the transmission rate for this type of traffic is adjusted depending on the bandwidth that remains following the allocation of resources to the real-time flows.

There are also other research projects on QoS in WMNs. DARE (Distributed Allocation of time slots for REal-time traffic) (Carlson *et al.* 2004) is an end-to-end resource reservation mechanism for multihop radio networks. This mechanism provides periodic reservation of time intervals for real-time flows. As for the other flows, they are transmitted using the conventional IEEE 802.11 mechanism based on the exchange of control messages between stations before transmission. In Wang and Mutka (2008), the authors present another QoS guarantee mechanism, QUOTA (quality-of-service aware rate fair allocation). This defines a framework that combines QoS guarantee and an equitable bandwidth distribution. The proposed framework prioritizes real-time flows by reserving bandwidth and distributes the remaining bandwidth evenly between elastic flows. A second framework proposed in Kim *et al.* (2013) combines the allocation of resources with the guarantee of QoS while also integrating an admission control mechanism. In this framework, the bandwidth requirement of each source node is verified while taking into account both interchannel interference as well as self-interferences. This latter type of interference considers that packets in the same flow may interfere with each other across a multihop route. In Li *et al.* (2016), the authors adopt a mechanism for adapting the width of the used channel by considering QoS when allocating resources in a WMN to improve connectivity. This is done by modeling this channel adaptation through an optimization problem in order to guarantee the resources required by the diverse traffic generated.

5.3.4. *Research projects on QoS in WMNs*

Certain research projects have studied and proposed solutions to guarantee satisfactory QoS in a WMN. Some of these projects are the CARMNET project (CARrier-grade delay-aware resource Management for wireless multihop/mesh NETworks) (CARMNET Project 2018), EU-MESH (Enhanced, Ubiquitous, and dependable broadband access using MESH network) (EU-MESH 2018), SMART-NET (SMART-antenna multimode wireless mesh NETwork) (Cordis 2018) and the WING project (wireless mesh network for next-generation Internet) (WING Project 2018).

5.3.4.1. *The CARMNET project*

The CARMNET project (CARMNET Project 2018) was a project funded by the Polish Swiss Research Program, which began in 2012 and ended in March 2015. The main objective of this project was to offer a new vision of WMNs. These networks can be used by a community of users who jointly share Internet access with telecom operators in order to allow the operators to extend coverage of their network without needing to enlarge their fixed infrastructure (Glabowski and Szwabe 2013). Thus, the proposed CARMNET framework is based on a highly reliable resource allocation system set up within a WMN, which may be incorporated into the infrastructure of any standard telecom operator.

To achieve the general objectives of this vision of CARMNET networks, several research subobjectives were set. These included developing a resource management algorithm based on the QoS delay parameter (Walkowiak *et al.* 2013), defining an improvement to the routing protocol used, managing QoS and mobility, and implementing an experimental environment (testbed) to evaluate the performance of the network.

5.3.4.2. *The EU-MESH project*

The EU-MESH project was a European FP7 project (7th Framework Program), which was launched in 2008 for a period of 30 months (EU-MESH 2018). The goal of this project was to develop, evaluate and test a system to construct reliable multichannel, multiradio WMNs with a QoS guarantee, while also providing ubiquitous and very high-speed Internet access. The subobjectives that the EU-MESH project set up were developing algorithms that would make it possible to combine mechanisms to access a channel with power control to reduce interference. The EU-MESH also aimed to develop an opportunistic routing protocol with end-to-end QoS guarantee for a scalable network. This project also proposed a framework (Tragos *et al.* 2010) that combined a mechanism for self-configuring WMN, an algorithm for allocating channels in a multiradio, multichannel environment and also proposed a QoS mechanism within this project. The self-configuring function is based on an improved version of DHCP to ensure the assignment of a unique IP address to each network interface in a multihop wireless network with multiple heterogeneous radio interfaces (Wi-Fi/WiMAX). The channel assignment function is based on an approach to optimize the mesh topology in terms of the throughput offered in this type of network. With respect to the proposed QoS mechanism, it is based on an approach to select the best gateway among available gateways in the mesh network for routing flows toward the Internet, while guaranteeing their QoS requirements. This approach makes it possible to prevent a bottleneck at a gateway by ensuring a better distribution of the traffic directed toward the Internet between the various gateways of the mesh network.

5.3.4.3. *The SMART-NET project*

The SMART-NET project (Cordis 2018) was also a European project under the FP7 program, launched in 2008 for a period of 3 years. This project was based on a wireless mesh architecture to define an infrastructure for broadband access that is not centralized on the Internet. This infrastructure makes it possible to overcome the limitation of conventional broadband networks where communications must pass through a central node, namely the base station. In order to achieve this objective, smart antennae, traditionally deployed in base stations, were used at each node of the mesh network. By combining the advantages of these antennae with mesh networks, SMART-NET offers a new architecture that can support a large number of simultaneous transmissions and new Internet applications. In addition, as a second objective, SMART-NET planned to reinforce this approach with new routing and scheduling protocols designed for heterogeneous WMNs (Wi-Fi/WiMAX). In this sense, an algorithm was proposed to monitor the bandwidth used by TCP flows with the aim of adapting it to the needs of VoIP traffic.

5.3.4.4. *The WING project*

The WING project (WING Project 2018) is a project that was funded by the Italian Ministry of Education, University and Research. The main objective of the project was to develop new algorithms and protocols to facilitate the role played by WMNs as standard architecture to access the Internet. Research in this project focused on several challenges including the consideration of multiple radio interfaces and channel allocation mechanisms, the self-configuration of nodes in a WING network, the guarantee of QoS and routing. The proposed QoS mechanism (Riggio *et al.* 2008) is based on a combination of service differentiation with a packet aggregation approach in a WMN using IEEE 802.11 technology. It is based on the use of four wait queues with different priorities and a weighted round robin (WRR) scheduler (Katevenis *et al.* 1991). A new packet aggregation technique is implemented for each wait queue. The mechanism makes it possible to determine the optimal size of the burst to be created based on the state of the link defined by the routing metric and the rate of channel use by the other neighboring nodes.

5.4. QoS-based routing for WMNs

The performance of WMNs is highly dependent on the routing protocols used. One of the research challenges in effectively meeting user needs was using a routing protocol to guarantee satisfactory QoS that is appropriate for different applications when choosing the routes used for routing the corresponding traffic in a WMN environment. Routing with guaranteed QoS remains an effective means of satisfying the requirements of different services in the specific field of WMNs.

5.4.1. *Routing requirements in WMNs*

In order to design a routing protocol that is well adapted to a WMN, a few properties must be taken into consideration. This routing protocol must, therefore, have the following characteristics:

– use of routing metrics that are better-performing than the basic "number of hops" metric (see section 5.4.2.1) so that the quality of links can be taken into account;

– a certain tolerance of errors: the routing protocol must be robust in the face of link failures. The routing protocol must be able to rapidly establish an alternative route;

– reliability: given its architecture, a WMN can offer multiple routes between each pair of nodes. These paths can be used by the routing protocol in the event of a failure on the primary route to reduce the packet loss ratio;

– load balancing: the routing protocol must be able to efficiently use the available resources, taking into account the constraints linked to bandwidth limitation, while avoiding routing data to congested areas of the network;

– scalability: the performance of the routing protocol must not be affected by an increase in network size;

– controlling overcharge introduced by its own control messages.

5.4.2. *Routing metrics in WMNs*

The routing metric represents the key parameter for the construction of routes by a routing protocol in a network. The metric can define the cost or weight of a link based on its quality, its capacity or even the diversity of channels at the level of the node in question. The metric for a route is determined by cumulating the metric of each of its links. In the following section, we will introduce the routing metrics that are used most often in WMNs.

5.4.2.1. *The hop count metric*

The hop count (HC) metric, or the number of hops, represents the distance between a source and destination in terms of nodes. This metric offers several advantages and as a result has been used for a long time as the main routing metric for wireless networks. This is a metric that is easy to implement, does not require an overly complex computation and does not load the network with additional control messages. Given the significant mobility of wireless nodes and the energy constraint, the HC metric is considered to be an efficient metric for conventional wireless networks. The selection of the shortest path makes it possible to effectively reduce the packet loss ratio due to the rupture of a link following the displacement or the disappearance of a node due to lack of energy. However, these constraints do not apply to a WMN characterized by a relatively static infrastructure with nodes connected to energy sources, thus allowing the deployment of more complex mechanisms at the level of mesh routers. In this sense, the new routing metrics for WMNs have been developed in order to offer routes with better performance (Guerin *et al.* 2007; Bharat *et al.* 2009) by considering the quality of the network links.

5.4.2.2. *The expected transmission count metric*

The expected transmission count (ETX) metric (De Couto *et al.* 2005) is one of the first routing metrics designed for a WMN. It estimates the number of transmissions and retransmissions necessary for a successful routing of a packet at a link. The success of a transmission is determined by the receipt of an acknowledgment (Ack) by the source. Thus, the computation of the ETX metric is based on two parameters. These are as follows: the probability that a packet is received by the recipient and the probability that an Ack has been received by the source. In effect, the ETX metric selects the routes based on the message loss ratio, thereby improving the reliability and the throughput of transmissions compared to the HC metric. This metric, however, does not consider the link load, available bandwidth or interferences. It also does not use any diversity there may be in channels at the node level.

5.4.2.3. *The expected transmission time metric*

The expected transmission time (ETT) metric (Draves *et al.* 2004) is an improvement on the ETX metric and takes into account the link bandwidth. It estimates the time required for a packet to be transmitted over a link according to its capacity. Capacity estimation is performed periodically using the "packet pair" technique (Keshav 1995). The ETT metric possesses the advantages of the ETX metric, while also improving the transmission throughput through the use of links with better bandwidth. However, the ETT metric does not consider the load at the level of a link, nor the interference that has come in. Thus, the weighted cumulative expected transmission time (WCETT) metric (Draves *et al.* 2004) was proposed as an improvement on the ETT metric to introduce the concept of channel diversity in order to contribute to the reduction of intraflow interference.

5.4.2.4. *The airtime link metric*

Airtime link (AL) is the default metric defined for IEEE 802.11s standard (IEEE 2011). This metric makes it possible to select routes based on the state of the radio links and reflects the delay required to transmit a frame over a particular link.

5.4.3. *QoS-based routing protocols in WMNs*

Several research projects have sought to integrate QoS into routing protocols in order to respond better to the requirements of different applications in a WMN.

5.4.3.1. *The wireless mesh routing protocol*

The wireless mesh routing (WMR) protocol (Xue and Ganz 2002) is one of the QoS-based routing solutions for WMNs. It offers different functionalities, allowing it to fulfill its routing task while providing an appropriate QoS in terms of bandwidth and delay. In its topology discovery phase, the nodes periodically broadcast their HELLO messages to recognize neighboring nodes. These messages contain, among other things, a field to store the distance between a node and the closest gateway in terms of hops. In this way, each node in the network has information on its neighborhood as well as on its nearest gateway for routing traffic to external networks. In addition to this, WMR also offers an admission control at the level of each node. This admission control is used during the route discovery phase to verify whether the available bandwidth is greater than required. WMR estimates bandwidth by measuring the bandwidth used by the node itself as well as by its neighbors. The delay metric is estimated by the round trip time (RTT), which defines the delay for transmission from the source to the destination and back to the source.

5.4.3.2. *The QUORUM protocol*

The QUORUM protocol (QUality Of service RoUting in wireless Mesh networks) (Kone *et al.* 2007) was designed to improve the WMR protocol by incorporating a new delay estimation mechanism. The end-to-end delay is estimated through messages called DUMMY RREP. These messages are used to emulate real data traffic since the control messages have characteristics different from those of a data packet. The use of DUMMY RREP messages makes it possible to obtain an estimate closer to the actual delay. Despite the precision provided by this method, it has drawbacks in terms of latency in setting up the route as well as overhead that may be brought in by its messages. In addition, the QUORUM protocol offers a technique for selecting the most robust routes. Each node is responsible for estimating the stability of the links with each of its neighbors. This is done by counting the number of HELLO messages received from each neighbor over a given time interval. If the number of messages is smaller than a given threshold, the protocol considers that this link is unstable and it will not be used for setting up a route.

5.4.3.3. *The interference and delay aware routing protocol*

The interference and delay aware routing (IDAR) protocol (Pal and Nasipuri 2011) was proposed for use in a WMN in order to remedy the problem of inaccurate estimation of the link quality using control messages that have different characteristics compared to data packets. IDAR uses a routing metric estimating the probability of success of transmissions (POS: probability of success of transmissions) as well as the delay, while taking into account any interferences that may occur. Nonetheless, this protocol adopts an estimation based on performance models for the experimentally obtained data packets. The IDAR protocol designers developed mathematical models from simulations carried out to define the factors that influence the quality of links. With regard to estimation of the delay, this depends chiefly on the delay in the wait queue and the delay in accessing the channel. Thus, following the evaluation of different simulation scenarios, a second-degree polynomial equation was defined to estimate the overall delay in the presence of interference. From these two models, defined for the estimation of the POS parameter and the delay, a new routing metric was developed to maximize the probability of success of the end-to-end transmission, while reducing the overall delay for the corresponding multihop route.

5.4.3.4. *The MARIA protocol*

The MARIA protocol (Mesh Admission control and QoS Routing with Interference Awareness) (Cheng *et al.* 2008) is a routing solution that guarantees QoS for multimedia flows in a WMN, while taking into account inter- and intraflow interferences. This solution uses a reactive routing protocol with a distributed admission control mechanism to guarantee bandwidth. The estimation of this QoS parameter at each node is carried out by taking into account nodes that are found in the zone of interference. MARIA is based on the concept of the conflict graph (Jain *et al.* 2003) to model the interference and subsequently determine the estimated value of the remaining bandwidth at a node.

5.4.3.5. *Other protocols*

Other routing protocols based on a verification of QoS during route discovery have also been proposed for WMNs. The QAODV protocol (QoS AODV) (Liu *et al.* 2012) verifies the same QoS parameters as the QUORUM protocol, namely bandwidth and delay, at each intermediate node. QAODV also considers the load rate for a route in order to estimate congestion at the nodes. Another variant of the AODV protocol, the R-AODV protocol (rate aware-AODV; Zhang *et al.* 2006), considers transmission time for a packet as a routing metric in a meshed Wi-Fi network with multiple flow rates. Several improvements have also been made to the reactive mode of the HWMP routing protocol for the IEEE 802.11s standard to guarantee a certain level of QoS for different types of applications. The Q-HWMP (QoS-HWMP) (Ben-Othman *et al.* 2010), for example, guarantees the best performance for real-time traffic by verifying a metric using end-to-end delay in conjunction with the default metric of the standard, the AL radio metric. The new Q-HWMP metric is calculated with respect to a packet by considering the time spent in the wait queue as well as the time required for a successful transmission including possible retransmission attempts. Another improvement in the HWMP relating to the QoS guarantee was proposed following the specification of the R-HWMP (reservation-based HWMP; Jung *et al.* 2013). This protocol defines an end-to-end bandwidth reservation mechanism, by adopting the MCCA (Mesh Coordination function coordinated Channel Access) access technique of the IEEE 802.11s standard. The MCCA technique defines a method of access to the channel that makes it possible to reserve time intervals for the transmissions from one node while also taking into consideration the communications of its neighbors. However, this reservation is limited to neighbors within two hops. The R-HWMP drew inspiration from the RSVP and introduced modifications to the conventional control messages and added the TSpec (traffic specification) and RSpec (reserve specification) RSVP fields to the PREQ and PREP messages, respectively. Thus, a resource verification is carried out at each intermediate node by receiving a PREQ message. After a PREP message has been received, the resources are reserved at each intermediate node using the MCCA technique.

The routing protocols that we described earlier rely on the use of the same better route for the different types of applications, thus causing congestion and overload problems on the established path. In Malgi and Gaikwad (2015), the authors improved the transmission quality of video traffic in a WMN based on IEEE 802.11 technology by differentiating it from data traffic. The proposed protocol was based on a cross-layer approach for the selection of the routing metric based on the information provided by the application layer. Thus, for data traffic, a route would be established based on the HC metric, while the ETX metric was used for video traffic. This dynamic adaptation of the routing metric to the type of traffic to be routed allows the discovery of a different better route for each application based on its QoS requirements. In this sense, a second reactive routing protocol was also defined in Yun *et al.* (2014). This would allow the use of five different metrics based on the type of traffic: the HC metric, the ETX metric, the ETT metric, a metric reflecting the available bandwidth and a metric reflecting the interference ratio. In Yamaguchi *et al.* (2016), the authors proposed routing with QoS guarantee based on the AODV and OLSR (optimized link state routing) protocols for a wireless mesh Wi-Fi environment. This routing technique made it possible to set up a route taking into account the network load while distinguishing between two classes of traffic, that is to say the real-time class and the non-real-time class.

A solution to enhance the performance of communications within an IoT environment based on WMNs was also proposed in Li *et al.* (2018). This environment is affected by the large amount of data generated, given the large number of objects. In this context, the authors proposed a clustering algorithm to better manage traffic by limiting control messages and considered a routing metric based on interference and load distribution.

5.5. HQMR: QoS-based hybrid routing protocol for mesh radio networks

In order to guarantee a satisfactory service level for the different flows based on their requirements within a WMN, we have proposed (Bargaoui *et al.* 2015) a QoS-based routing framework that defines a hybrid routing protocol for a wireless mesh architecture, i.e. the HQMR protocol (hybrid QoS mesh routing).

5.5.1. *Description of the HQMR protocol*

5.5.1.1. *General architecture of the framework*

Within our QoS-based routing framework, we define a wireless mesh infrastructure formed by a set of fixed mesh nodes interconnected with each other. As shown in Figure 5.2, the wireless mesh infrastructure provides the connection between the mesh nodes and external networks, including the Internet, using the MG.

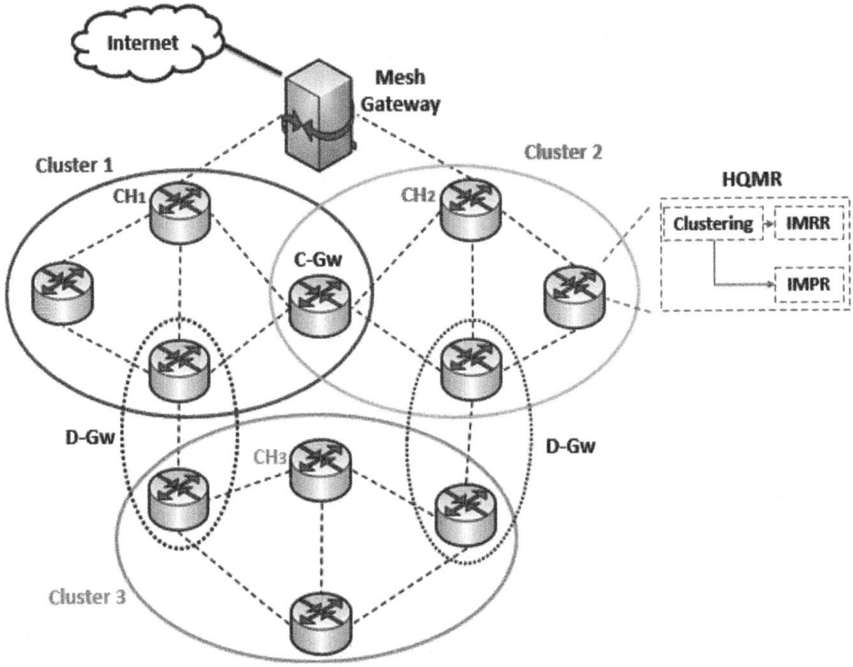

Figure 5.2. *General architecture of the framework. For a color version of this figure, see www.iste.co.uk/mbarek/service.zip*

In order to guarantee a satisfactory level of QoS within the infrastructure, we implement a QoS-based hybrid routing protocol, i.e. HQMR, at each node of the infrastructure including the MG.

5.5.1.2. *Design principle for the HQMR protocol*

The HQMR routing protocol (Bargaoui *et al.* 2015) is a hybrid QoS-based routing protocol. This protocol aims to ensure the correct routing of flows with different QoS constraints for communications within the wireless mesh infrastructure as well as those to external networks. To this end, we define three classes of service (see section 5.5.1.3) in order to meet the QoS requirements of different types of applications. The design of the HQMR protocol is mainly based on three functional blocks that are defined as follows:

– A clustering block: dividing the network into a set of clusters makes it possible to effectively reduce the load of control messages since their exchange is limited to a cluster. Clustering, thus, improves the performance of the routing protocol. In this sense, we propose a clustering algorithm (Bargaoui *et al.* 2014), which we associate with routing functionalities within the HQMR protocol. This algorithm divides the infrastructure into a set of single-hop clusters (see Figure 5.2), each managed by a node called the cluster head (CH). Two types of gateways are then defined, called the cluster gateway (C-Gw) and the distributed gateway (D-Gw), and they maintain connectivity between the different clusters. This division into clusters makes it possible to minimize the sizes of the routing tables and to facilitate the establishment of the routes because of intercluster communications.

– A first routing subprotocol, namely IMRR (intra-mesh infrastructure reactive routing): this is a reactive routing protocol, which ensures a QoS guarantee for intra-infrastructure communications. IMRR establishes a different route with a satisfactory QoS for each defined class of service. Thus, these different routes will allow a better adaptation to the requirements of the different classes of traffic and a reduction of the disturbances caused by flows without constraints to flows with QoS constraints.

– A second routing subprotocol, namely IMPR (inter-mesh infrastructure proactive routing): this is a proactive routing protocol, which makes it possible to offer a satisfactory level of QoS for communications to external networks, including the Internet. IMPR is a tree-based multipath routing protocol, in which the MG gateway is configured as the root. It guarantees the construction of three partially disjointed QoS-based routing trees, so that each tree is used to route traffic for a specific class of service.

5.5.1.3. *QoS guarantee with HQMR*

HQMR provides a QoS guarantee for intra-infrastructure communications, as well as for inter-infrastructure communications. This guarantee is achieved by differentiating three classes of service, namely the real-time class, the streaming class and the BE class. The real-time class covers interactive real-time applications such as VoIP and videoconferencing. These applications are very sensitive to delay and jitter. The streaming class represents streaming applications such as VoD and IPTV. These applications are more sensitive to jitter. As for the BE class, it is made up of different applications generating elastic traffic with few QoS constraints, such as file transfers.

In order to meet the requirements of different applications, the HQMR protocol, through its IMRR and IMPR subprotocols, takes into account three QoS metrics: available bandwidth, delay and jitter. Depending on the class of traffic concerned, a combination of these metrics will be used to define the routes to prioritize during the routing of data.

5.5.2. *How the HQMR protocol works*

5.5.2.1. *IMRR: reactive QoS-based subprotocol*

The IMRR protocol (Bargaoui *et al.* 2015) is the reactive subprotocol of the HQMR routing protocol developed to ensure a QoS guarantee for intra-infrastructure communications. It is used to route traffic between two mesh nodes in the infrastructure, whether or not they belong to the same cluster, while ensuring a satisfactory level of QoS with respect to the class of traffic to be forwarded.

The IMRR protocol is based on an exchange of RREQ (route request) and RREP (route reply) messages for the process of establishing routes between the source and the destination. A mesh source, wishing to communicate with a mesh node within the infrastructure, broadcasts an RREQ message to find a suitable route for the traffic class to be transmitted. In order to reduce the load introduced by the protocol, this broadcast is limited to CH nodes, clustering gateways (C-Gw and/or D-Gw) and MGs. However, in order to guarantee the discovery of the best route in terms of QoS, the receiving nodes redirect all the received RREQ messages (duplicated or not) to the recipient if these messages verify the required QoS. Once an RREQ message, with a satisfactory QoS, is received by the recipient, the latter responds with an RREP message to the source for validation of the corresponding route.

The IMRR protocol guarantees QoS by providing each class of service with a different route depending on the constraints of the corresponding type of traffic. In effect, IMRR is based on the use of different routes between a source and a destination for flows belonging to different classes of service, so as to reduce the disruption of flows with QoS constraints by those without constraints. Thus, a route discovery procedure is launched separately for each class of service, even if a valid route to the same recipient exists in the routing table for a different class of service than that of the traffic to be carried. The recipient node then responds differently to a RREQ message depending on the associated type of traffic. For real-time class traffic, the recipient responds to the first RREQ message received that verifies the requested QoS. For streaming class traffic, the recipient responds to the second RREQ message received with a satisfactory QoS, given the route discovered by the first RREQ message could be used by traffic of the first real-time class. Finally, for BE traffic, the recipient responds to the third RREQ message received, so as to produce less interference with the other streams related to the classes with QoS constraints (i.e. real-time and streaming).

The verification of QoS when building the IMRR protocol routes is based on multicriteria metric composed of bandwidth, delay and jitter. Thus, to best meet the constraints of each class, this multicriteria metric is adapted to each class. The IMRR thus verifies the overall metric for the real-time class, only verifies the bandwidth and the jitter for the streaming class and ensures a selection of routes without constraints for the BE class.

5.5.2.2. *IMPR: QoS-based proactive subprotocol*

The IMPR protocol (Bargaoui *et al.* 2015) is the proactive subprotocol of the HQMR global routing protocol. It was designed to provide a QoS guarantee for inter-infrastructure communications, specifically toward the Internet. IMPR is a tree-based routing protocol, whose root is the MG and whose branches are formed only by the different CH nodes, the C-Gw gateways and the D-Gw gateways of the wireless HQMR mesh infrastructure. In addition, IMPR adopts the concept of multipath routing to define three different and partially disjointed routes between each mesh node and the root so as to allocate each of these routes for the delivery of a specific class of service. This procedure is based on an exchange of three control messages, namely the RANN message (root announcement), the PREQ message (path request) and the PREP message (path reply). As with the IMRR protocol and in order to reduce the additional load introduced, the various control messages of the IMPR protocol are only processed by the CH nodes and/or the C-Gw nodes and/or the D-Gw nodes and/or the MG gateway, called leaf nodes. To summarize, the IMPR protocol allows the construction of three partially disjointed routing trees based on QoS, which have the MG gateway as a common root. Each tree is used for the routing of a defined class of service, thus differentiating between the real-time tree, the streaming tree and the BE tree.

The working of the IMPR protocol is based on two principal phases, namely the route storage phase and the tree-construction phase. The first phase is executed at each mesh node participating in IMPR routing to store the maximum possible number of routes to the root in a RouteCache table using RANN messages. The goal of this phase is to store the different possible routes to the root, with the corresponding QoS metrics, in the RouteCache table of each leaf node. This is done before proceeding with the construction of the three routing trees. Thus, in the context of this first storage phase, the root starts by multicasting the RANN control message to all its neighbors to announce its presence. Each leaf node that receives this message records the route to the root and retransmits it in multicast so that it is received by all the nodes of the mesh infrastructure.

Once this first phase is complete, each node starts the process of building the routing trees. Thus, a leaf node begins by choosing a potential route for the first tree (i.e. the real-time tree) in its RouteCache table by executing a route selection algorithm. This potential route is then validated as being one of the branches of the tree by an exchange of PREQ/PREP messages with the root. Indeed, this exchange of control messages, between each leaf node and the root, makes it possible to ensure that each node of the same branch uses the same route to the root. Thus, we avoid the fact that a node can have more than one route to the root for a given tree. Once the route for the routing tree under construction is validated following the reception of a PREP message, an IMPR node updates the state of this route in the RouteCache table as being that chosen for the definition of a given tree. The node subsequently launches the procedure for building the second tree as well as the procedure for the third using the same approach.

The route selection algorithm used during the second phase is defined so as to allow each IMPR node to choose, from its RouteCache table, a potential path to validate for each routing tree while meeting the QoS requirements of each of the defined service classes. For the first tree, corresponding to the real-time tree, IMPR allows you to choose the best path in the RouteCache table, in terms of delay and jitter with a satisfactory level of bandwidth. For the streaming tree, IMPR chooses a path that is partially disconnected from the validated path for the first tree, in order to reduce congestion problems, and which also offers a satisfactory level of QoS in terms of jitter. Finally, among the remaining routes, IMPR makes it possible to select the best route for the third tree in terms of disjunction with respect to the first two paths already validated.

5.5.3. *Validation of the HQMR protocol*

In order to validate the HQMR routing protocol, we developed its source code using the ns-3 network simulator. Subsequently, we carried out various simulation scenarios to evaluate its performance. In this section, we will present some of the simulation results for evaluating the performance of real-time traffic using routes established by HQMR in a WMN.

5.5.3.1. *Simulation environment*

For the performance evaluation scenario, we considered a wireless mesh architecture formed by a maximum of 50 fixed mesh nodes, distributed randomly. The algorithm of the HQMR protocol was implemented at the level of each mesh node. The bandwidth of all nodes on the network was 6 Mbit/sec and the radio range was 150 m. The duration of the simulation was fixed at 200 sec and the results obtained corresponded to the average values of 20 iterations.

To demonstrate the efficiency of the HQMR protocol during the routing of real-time interactive applications (real-time class), we evaluated the performance of a VoIP application in terms of end-to-end delay and jitter, given the sensitivity of this type of application to these two QoS parameters. These performance parameters were evaluated by varying the number of nodes from 6 to 50 in the wireless mesh topology. For the test scenario, we used a traffic model corresponding to the G711 codec (International Telecommunications Union 2018b) to simulate VoIP traffic. This model generated 50 packets/sec, and the size of each packet was 160 bytes (equal to a flow rate of 64 Kbit/sec). However, in order to overload certain links of the infrastructure, other source nodes transmitted CBR traffic with a rate of 512 Kbit/sec and packets whose size was 512 bytes.

5.5.3.2. *Evaluating the IMRR subprotocol*

To demonstrate that IMRR can enhance performance, we carried out simulations to compare performances (in terms of delay and jitter) of VoIP applications using routes discovered by the IMRR routing protocol with VoIP applications using routes discovered by the AODV protocol, in a random wireless mesh architecture. The generated traffic was sent from a source node to the farthest destination node in the infrastructure.

The comparative results of this evaluation are given in Figure 5.3. We observe a significant difference in the delay and jitter parameters offered by the route selected by each of the two protocols. For example, we note that in an architecture of 30 mesh nodes there is a delay of 14.39 msec for a VoIP application using a route discovered by the IMRR protocol, as against a delay of 28.48 msec for a route discovered by AODV. Indeed, the AODV protocol uses only the HC metric for the discovery of routes and does not take into account the state of the links. However, the IMRR protocol verifies the QoS metrics, namely the bandwidth, the delay and the jitter at each intermediate node during the route discovery phase in order to respect the constraints relating to a VoIP application. This procedure guarantees the IMRR protocol the discovery of a route with the best QoS, allowing the proper functioning of this VoIP application.

We also carried out other simulations to compare the IMRR protocol with other QoS-based protocols, namely the MARIA and the IDAR protocols. We evaluated the delay and jitter parameters according to the number of streams introduced into the network.

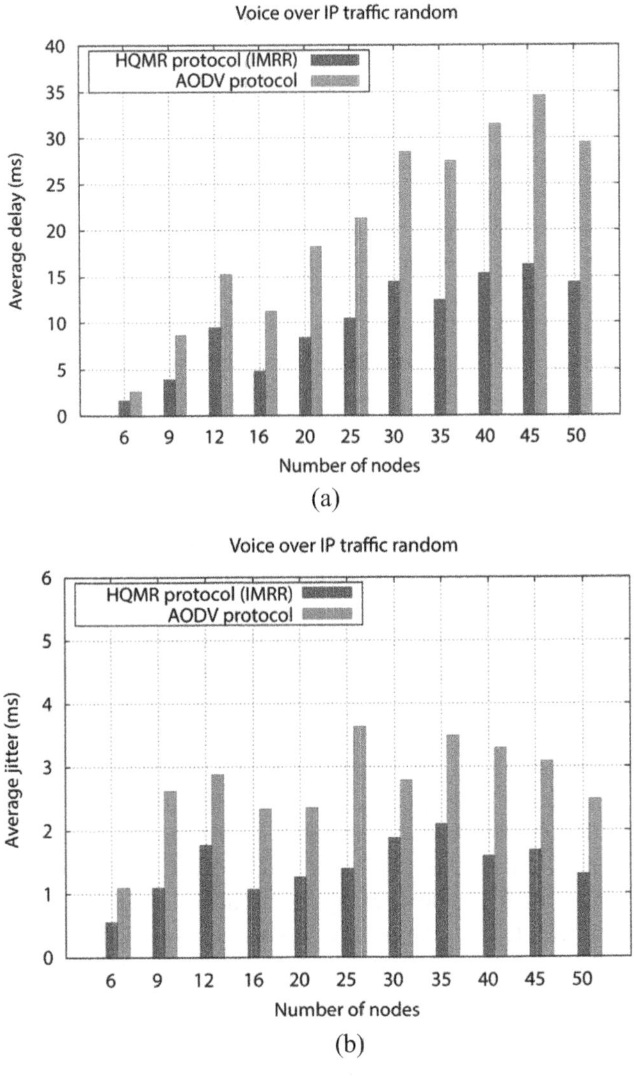

Figure 5.3. *Evaluating (a) average end-to-end delay and (b) jitter for a VoIP application based on the number of nodes. For a color version of this figure, see www.iste.co.uk/mbarek/service.zip*

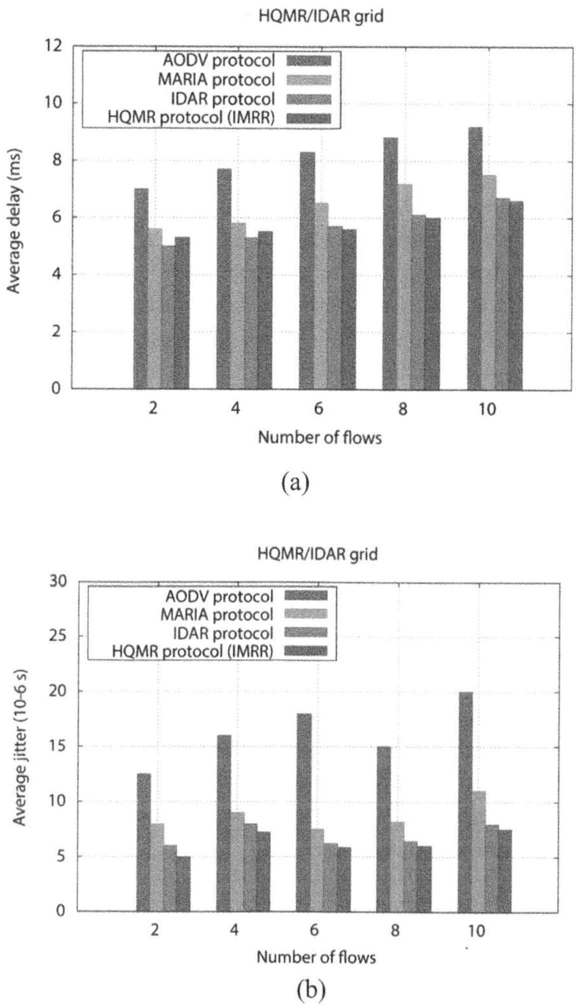

Figure 5.4. *Performances of the IMRR protocol versus MARIA and IDAR. Evaluation (a) of the average end-to-end delay and (b) of the average jitter. For a color version of this figure, see www.iste.co.uk/mbarek/service.zip*

It can be seen in Figure 5.4 that the MARIA protocol offered better performance values than the AODV protocol in terms of end-to-end delay and jitter observed for the traffic in the simulation. This is explained by the fact that the shortest route discovered by AODV does not necessarily have better QoS. The MARIA protocol considers the quality of the link in terms of interference when discovering the route to use. However, the IDAR and IMRR protocols both offer better performance values than the MARIA protocol in terms of end-to-end delay and jitter when routing traffic. In fact, MARIA selects a route based on the residual bandwidth and the interference ratio as routing metric, without considering the delay at the link level, whereas the IDAR protocol is based on a metric estimating the probability of success of transmissions and the delay for the selection of the route. However, unlike the MARIA protocol, the IMRR protocol selects the best route, taking into account the QoS parameters of delay and jitter. Further, it can be seen that the IMRR protocol has slightly better values in terms of jitter as compared to the IDAR protocol, given that IDAR only considers the delay parameter in the selection of its routes. Thus, the IMRR protocol offers almost the same performance as the IDAR protocol, since all the source nodes are configured to generate the same type of flow. Consequently, in this case, the IMRR protocol uses a single route, the best in terms of delay and jitter, to route data from the same type of flow. This scenario does not allow IMRR to take advantage of its differentiation mechanism to offer different routes to applications with different QoS constraints belonging to the classes defined by HQMR and thus avoid the problems caused by the use of the same route.

5.5.3.3. *Evaluating the IMPR subprotocol*

To demonstrate the effectiveness of the IMPR routing protocol when routing real-time interactive applications to outside networks, we evaluate the performance of a VoIP application in terms of end-to-end delay and jitter. We carried out various simulations to compare the performance of traffic routed using routes defined by the IMPR routing protocol, the HWMP protocol and the OLSR protocol in a random wireless mesh architecture. The real-time traffic generated was sent from a mesh node to the external network through the MG.

The comparative results of this evaluation are given in Figure 5.5. From these results, it can be seen that the routes defined by the IMPR protocol trees guarantee the VoIP traffic a delay and jitter that are equal to half the values of its QoS parameters when using the HWMP tree or against the OLSR proactive route. These performance values obtained because of the IMPR routing protocol allow a better guarantee of QoS for applications belonging to the real-time class in the context of inter-infrastructure communications. Indeed, the OLSR protocol is only based on the HC metric for the selection of its routes without any verification of the quality of the links. In addition, the HWMP protocol ensures the construction of a single routing tree for the different types of traffic to be routed based on the radio metric AL. Thus, a real-time flow can be disrupted by other flows without QoS constraints using the same route specified by OLSR or HWMP. However, the IMPR protocol offers three partially disjointed routing trees, which make it possible to offer a satisfactory level of QoS for each traffic class according to its requirements.

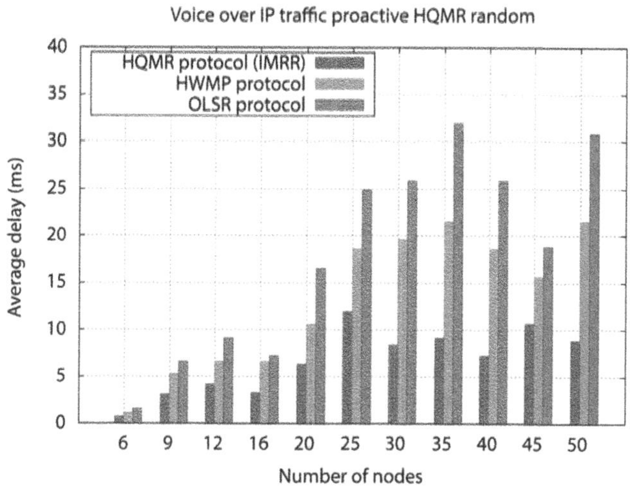

(a)

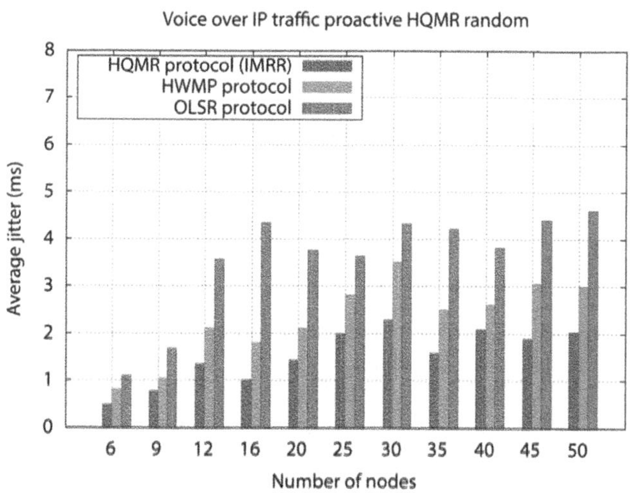

(b)

Figure 5.5. *Evaluation of (a) average end-to-end delay and (b) average jitter for a VoIP application based on the number of nodes. For a color version of this figure, see www.iste.co.uk/mbarek/service.zip*

5.6. Conclusion

In this chapter, we studied different concepts related to WMNs, and in particular QoS management, which show how important this type of network is with the emergence of new fields of application. We first presented the most important architecture, characteristics and standardization activities in networks of this kind. In this sense, we have presented the two most important standards of WMNs based on different wireless technologies, in particular the IEEE 802.16j standard and the IEEE 802.11s standard. We then defined the general concept of QoS and the various mechanisms proposed in this context for WMNs. We also described research projects studying QoS in WMNs. Then, we introduced the concept of QoS-based routing as one of the existing solutions to guarantee a satisfactory level of QoS when routing data within a WMN. We detailed various routing metrics based on QoS making it possible to improve performance during the routing of real-time and multimedia applications in a WMN before describing some QoS-based routing protocols for this type of network. Finally, we presented a routing protocol (HQMR) proposed in order to guarantee QoS for intra- and inter-infrastructure communications in a WMN through the specification of the two subprotocols, IMRR and IMPR. A performance evaluation using the ns-3 simulator shows the efficiency of the HQMR protocol when routing real-time flows in a wireless mesh environment. HQMR offers these types of streams better delays and jitters compared to routing protocols such as HWMP, MARIA and IDAR.

5.7. References

Akyildiz, I.F. and Wang, X. (2005). A survey on wireless mesh networks. *IEEE Communications Magazine*, 43(9), S23–S30.

Akyildiz, I.F., Wang, X., and Wang, W. (2005). Wireless mesh networks: A survey. *Computer Networks*, 47(4), 445–487.

Andrés, G.R.C. (2016). CleanWiFi: The wireless network for air quality monitoring, community Internet access and environmental education in smart cities. *ITU Kaleidoscope: ICTs for a Sustainable World*, 1–6.

Aruba (2018a). AirMesh outdoor wireless mesh networks [Online]. Available at: http://www.arubanetworks.com/products/networking/outdoor-mesh/ [Accessed 5 January 2020].

Aruba (2018b). A Hewlett Packard Enterprise Company. Solutions pour le secteur des services de santé [Online]. Available at: http://www.arubanetworks.com/fr/solutions/services-de-sante/ [Accessed 5 January 2020].

Bargaoui, H., Mbarek, N., Togni, O., and Frikha, M. (2014). Hybrid QoS based routing for IEEE 802.16j mesh infrastructure. *The 10th Advanced International Conference on Telecommunications*, IEEE, Paris, 110–118.

Bargaoui, H., Mbarek, N., Togni, O., and Frikha, M. (2015). Hybrid QoS based routing protocol for inter and intra wireless mesh infrastructure communications. *Wireless Networks*, 22(7), 1–20.

Ben-Othman, J., Mokdad, L., and Cheikh, M.O. (2010). Q-HWMP: Improving end-to-end QoS for 802.11s based mesh networks. *Global Telecommunications Conference*, IEEE, Miami, 1–6.

Bharat, K.A., Vineeth, K., and Kiran, D. (2009). A survey: Routing metrics for wireless mesh networks. Document [Online]. Available at: https://www.slideshare.net/networksguy/a-survey-routing-metrics-for-wireless-mesh-networks [Accessed 5 January 2020].

Blackbox (2018). SmartPath enterprise wireless [Online]. Available at: https://www.blackbox.fr/fr-fr/page/21967 [Accessed 5 January 2020].

Carlson, E., Karl, H., Carlson, E., Holger, K., Wolisz, A., and Prehofer, C. (2004). Distributed allocation of time slots for real-time traffic in a wireless multi-hop network. *Proccedings European Wireless*, IEEE, Barcelona.

CARMNET Project (2018). CARMNET [Online]. Available at: http://carmnet.put.poznan.pl [Accessed 5 January 2020].

Cheng, X., Mohapatra, P., Lee, S., and Banerjee, S. (2008). MARIA: Interference-aware admission control and QoS routing in wireless mesh networks. *International Conference on Communications*, IEEE, Beijing, 2865–2870.

Cordis (2018). Smart-net [Online]. Available at: http://cordis.europa.eu/projects/rcn/86666_fr.html [Accessed 5 January 2020].

Crawley, E., Sandick, H., Nair, R., and Rajagopalan, B. (1998). A framework for QoS-based routing in the Internet. Working document [Online]. Available at: http://tools.ietf.org/html /rfc2386 [Accessed 5 January 2020].

De Couto, D.S.J., Aguayo, D., Bicket, J., and Morris, R. (2005). A high-throughput path metric for multi-hop wireless routing. *Wireless Networks*, 11(4), 419–434.

Dhurandher, S.K., Woungang, I., Obaidat, M.S., Kumar, K., Joshi, M., and Verma, M. (2015). A distributed adaptive admission control scheme for multimedia wireless mesh networks. *IEEE Systems Journal*, 9(2), 595–604.

Draves, R., Padhye, J., and Zill, B. (2004). Routing in multi-radio, multi-hop wireless mesh networks. *Proceedings of the 10th Annual International Conference on Mobile Computing and Networking*, Mobicom, New York, 114–128.

EU-MESH. (2018). EU-Mesh: Enhanced, ubiquitous and dependable broadband access using MESH networks [Online]. Available at: http://www.eu-mesh.eu/ [Accessed 5 January 2020].

Firetide (2018a). Firetide for public safety [Online]. Available at: http://www.firetide.com/solutions/public-safety/ [Accessed 5 January 2020].

Firetide (2018b). Firetide for transportation [Online]. Available at: http://www.firetide.com/solutions/transportation/ [Accessed 5 January 2020].

Glabowski, M. and Szwabe, A. (2013). Carrier-grade Internet access sharing in wireless mesh networks: The vision of the carmnet project. *Proceedings of the 10th Advanced International Conference on Telecommunications*, AICT, Rome, 113–116.

Google Store (2018). Google Wi-Fi [Online]. Available at: https://store.google.com/fr/product/google_wifi [Accessed 5 January 2020].

Guerin, J., Portmann, M., and Pirzada, A. (2007). Routing metrics for multi-radio wireless mesh networks. *Telecommunication Networks and Applications Conference*, ATNAC, Christchurch, 343–348.

IEEE (2009). Standard for local and metropolitan area networks part 16: Air interface for broadband wireless access systems amendment 1: Multihop relay specification. *IEEE Std 802.16j-2009*, 1–290.

IEEE (2011). Standard for Information Technology – Telecommunications and information exchange between systems – Local and metropolitan area networks – Specific requirements – Part 11: Wireless LAN medium access control (MAC) and physical layer (PHY) specifications – Amendment 10: Mesh networking. *IEEE*, 1–372.

IEEE (2012a). Standard for Information Technology – Telecommunications and information exchange between systems – Local and metropolitan area networks – Specific requirements – Part 11: Wireless LAN medium access control (MAC) and physical layer (PHY) specifications. *IEEE Std 802.11-2012*, 1–2793.

IEEE (2012b). Standard for air interface for broadband wireless access systems. *IEEE Std 802.16-2012*, 1–2542.

International Telecommunications Union (2018a). Recommandation E.802 (02/07) : cadre et méthodes de détermination et d'application des paramètres de qualité de service [Online]. Available at: https://www.itu.int/rec/T-REC-E.802-200702-I/fr [Accessed 5 January 2020].

International Telecommunications Union (2018b). G.711 [Online]. Available at: http://www.itu.int/rec/T-REC-G.711/ [Accessed 5 January 2020].

Jain, K., Padhye, J., Padmanabhan, V., and Qiu, L. (2003). Impact of interference on multi-hop wireless network performance. *Proceedings of the 9th Annual International Conference on Mobile Computing and Networking*, ACM, New York.

Jung, W.J., Min, S.H., Kim, B.G., Choi, H.S., Lee, J.Y., and Kim, B.C. (2013). R-HWMP: Reservation-based HWMP supporting end-to-end QoS in wireless mesh networks. *International Conference on Information Networking*, ICOIN, Bangkok, 385–390.

Katevenis, M., Sidiropoulos, S., and Courcoubetis, C. (1991). Weighted round-robin cell multiplexing in a general-purpose ATM switch chip. *IEEE Journal on Selected Areas in Communications*, 8, 1265–1279.

Keshav, S. (1995). A control-theoretic approach to flow control. *Proceedings of the Conference on Communications Architecture & Protocols*, ACM, New York, 21(4), 3–15.

Khasawneh, F.A., Benmimoune, A., and Kadoch, M. (2016). Joint routing and admission control in wireless mesh network. *International Journal of Communications, Network and System Sciences*, 9(8) 311–325.

Kiani, A.K., Ali, R.F., and Rashid, U. (2015). Energy-load aware routing metric for hybrid wireless mesh networks. *Vehicular Technology Conference (VTC Spring)*, IEEE, Glasgow, 1–5.

Kim, T., Yang, Y., Hou, J.C., and Krishnamurthy, S.V. (2013). Resource allocation for QoS support in wireless mesh networks. *IEEE Transactions on Wireless Communications*, 5, 2046–2054.

Kone, V., Das, S., Zhao, B., and Zheng, H. (2007). QUORUM: Quality of service in wireless mesh networks. *Mobile Networks and Applications*, 12(5), 358–369.

Li, H., Zhang, J., Hong, Q., Zheng, H., Wang, Y., and Zhang, J. (2016). QoS-aware channel-width adaptation in wireless mesh networks. *International Conference on Communications*, IEEE, Kuala Lumpur, 1–6.

Li, J., Silva, B.N., Diyan, M., Cao, Z., and Han, K. (2018). A clustering based routing algorithm in IoT aware wireless mesh networks. *Sustainable Cities and Society*, 40, 657–666.

Liu, L., Zhu, L., Lin, L., and Wu, Q. (2012). Improvement of AODV routing protocol with QoS support in wireless mesh networks. *Physics Procedia*, 25, 1133–1140.

Liu, Y., Tong, K.F., Qiu, X., Liu, Y., and Ding, X. (2017). Wireless mesh networks in IoT networks. *International Workshop on Electromagnetics: Applications and Student Innovation Competition*, London, 183–185.

Malandra, F. and Sansò, B. (2017). Performance analysis of large scale RF-mesh networks for smart cities and IoT. *International Conference on Smart Grid Communications (SmartGridComm)*, IEEE, Dresden, 13–18.

Malgi, M.A. and Gaikwad, G.N. (2015). A study on QoS enhancement of MPEG-4 video transmission over wireless mesh network. *International Conference on Pervasive Computing and Communications*, IEEE, Pune, 1–5.

Naveen, T.H. and Vasanth, G. (2017). Qualitative study of existing research techniques on wireless mesh network. *International Journal of Advanced Computer Science and Applications*, 8(3), 49–57.

Nunez-Martinez, J. and Mangues-Bafalluy, J. (2010). A survey on routing protocols that really exploit wireless mesh network features. *Journal of Communications*, 5(3), 211–231.

Open Garden (2018). Start something [Online]. Available at: http://www.open garden.com/firechat.html [Accessed 5 January 2020].

Open-Mesh (2018). Cloud-managed wireless networks [Online]. Available at: http:/www.open-mesh.com/ [Accessed 5 January 2020].

Pal, A. and Nasipuri, A. (2011). A quality based routing protocol for wireless mesh networks. *Pervasive and Mobile Computing*, 7(5), 611–626.

Pirzada, A.A., Wishart, R., Portmann, M., and Indulska, J. (2009). ALARM: An adaptive load-aware routing metric for hybrid wireless mesh networks. *Proceedings of the 32nd Australasian Conference on Computer Science*, Sydney, 37–46.

Plageras, A.P., Psannis, K.E., Ishibashi, Y., and Kim, B.G. (2016). IoT-based surveillance system for ubiquitous healthcare. *42nd Annual Conference of the Industrial Electronics Society*, IEEE, Florence, 6226–6230.

Portmann, M. and Pirzada, A.A. (2008). Wireless mesh networks for public safety and crisis management applications. *IEEE Internet Computing*, 1, 18–25.

Prabakar, D., Pradeep, J., Elango, G.A., and Lakshmy, N. (2012). Performance analysis of wireless mesh network in hospital environment. *Communication and Information Systems*, 28–33.

Purri, S., Choudhury, T., Kashyap, N., and Kumar, P. (2017). Specialization of IoT applications in health care industries. *International Conference on Big Data Analytics and Computational Intelligence*, IEEE, Chirala, 252–256.

Rao, A.N. and Rao, C.D.V.S. (2015). Way-point multicast routing framework for improving QoS in hybrid wireless mesh networks. *Wireless Network*, 22, 1–14.

Riggio, R., Miorandi, D., De Pellegrini, F., Granelli, F., and Chlamtac, I. (2008). A traffic aggregation and differentiation scheme for enhanced QoS in IEEE 802.11-based wireless mesh networks. *Computer Communications*, 31(7), 1290–1300.

Rodas-Andrade, M., Zea-Polo, C., and Inga-Ortega, J. (2017). Optimum design and dimensioning model of a Mesh-WiFi network for emergency services in protected areas. *IEEE 2nd Ecuador Technical Chapters Meeting (ETCM)*, Salinas, 1–6.

Ruckus Wireless (2013). SmartMesh networking [Online]. Available at: https://www.ruckuswireless.com/technology/smartmesh [Accessed 5 January 2020].

Shen, Q., Fang, X., Li, P., and Fang, Y. (2008). Admission control for providing QoS in wireless mesh networks. *International Conference on Communications*, IEEE, Beijing, 2910–2914.

Siddiqui, M.S., Amin, S.O., and Hong, C.S. (2008). An efficient mechanism for network management in wireless mesh network. *10th International Conference on Advanced Communication Technology*, IEEE, Gangwon-Do, 301–305.

Strix (2018a). Solution for safety [Online]. Available at: http://www.strixsystems.com/publicsafety.aspx [Accessed 5 January 2020].

Strix (2018b). Solution for transportation [Online]. Available at: http://www.strixsystems.com/transportation.aspx [Accessed 5 January 2020].

Tragos, E.Z., Brunoy, R., Ancillottiz, E., Grochlaz, K., and Siris, V.A. (2010). Automatically configured, optimised and QoS aware wireless mesh networks. *IEEE 21st International Symposium on Personal Indoor and Mobile Radio Communications*, Istanbul, 2081–2086.

Walkowiak, P., Urbanski, M., Poszwa, M., and Szalski, R. (2013). Flow classification in delay-aware NUM-oriented wireless mesh networks. *The 6th International Conference on Advances in Mesh Networks*, Barcelona, 33–38.

Wang, B. and Mutka, M. (2008). QoS-aware fair rate allocation in wireless mesh networks. *Computer Communications*, 31(7), 1276–1289.

WING Project (2018). WING Project [Online]. Available at: http://www.wing-project.org/doku. php [Accessed 5 January 2020].

Xue, Q. and Ganz, A. (2002). QoS routing for mesh-based wireless LANs. *International Journal of Wireless Information Networks*, 9(3), 179–190.

Yamaguchi, K., Mukaiyama, K., Akiyama, T., Kobayashi, T., and Matsue, H. (2016). Performance analysis of routing methods based on OLSR and AODV with traffic load balancing and QoS for Wi-Fi mesh network. *International Conference on Information Networking*, Kota Kinabalu, 114–119.

Yun, J., Han, J., Seong, G., Cho, W., Seo, J., Khan, M., Kim, B., Park, G., and Han, K. (2014). Self-organized multi-metric routing for QoS in wireless mesh networks. *International Conference on Information Networking*, Siem Reap, 160–163.

Zhai, H., Chen, X., and Fang, Y. (2004). A call admission and rate control scheme for multimedia support over IEEE 802.11 wireless LANs. *International Conference on Quality of Service in Heterogeneous Wired/Wireless Networks*, Dallas, 76–83.

Zhang, Y. (2006). *Wireless Mesh Networking: Architectures, Protocols and Standards*. Auerbach Publications, London.

Zhang, Y., Wei, Y., Song, M., and Song, J. (2006). R-AODV: Rate aware routing protocol for WiFi mesh networks. *International Conference on Wireless, Mobile and Multimedia Networks*, Hangzhou, 1–4.

6

Blockchain Based Authentication and Trust Management in Decentralized Networks

Axel MOINET[1] and Benoît DARTIES[2]

[1] *LIB University of Bourgogne Franche-Comté, Dijon, France*

[2] *LIRMM, University of Montpellier and CNRS, Montpellier, France*

6.1. Introduction

In the Internet of Things (IoT), sensor networks form the basis for interactions with the environment and are seeing accelerated development. In 2017, Gartner's projections estimated that the number of connected objects in the world was 8.4 billion objects (Gartner 2017). The constraints on nodes that make up these networks and their mobility complicate communications and prevent the use of algorithms that are deployed in conventional wireless networks: the embedded platforms under consideration have highly restricted resources in terms of computation capacity and memory, as well as limited energy resources (Akyildiz *et al.* 2002). Consequently, the algorithms and architecture used must include energy consumption as a strict constraint. Further, maintenance operations and human intervention must be as infrequent as possible (Karl and Willig 2005): systems may be deployed in zones that are geographically difficult to access and it is not possible to plan for recurrent operations following multiple malfunctions related to the large number of components deployed. At the same time, we must facilitate use and deployment and allow applications to participate in the environment while also optimizing the available resources.

Service Level Management in Emerging Environments,
coordinated by Nader MBAREK.

These networks suffer today from the fact that security measures are not adapted to their multiple constraints. Authentication and trust management solutions such as Pretty Good Privacy (PGP) only partially respond to their needs and constraints. On the one hand, the autonomy and cooperation constraints among nodes that are required for guaranteeing consistency in the network require a distributed solution. On the other hand, the constraints around energy consumption and the low computing power of the nodes require the use of algorithms with low complexity (Zhang *et al.* 2014). At the same time, recrudescence of IT attacks (both against and using the network of sensors and connected objects) and the incompatibility between existing solutions and the constraints on this network require an enquiry into new models, algorithms and structures for data dedicated to security.

Challenges around authentication and trust management may be considered two faces of the same problem. Authentication makes it possible to acquire information about the identity of the different actors present, whereas trust provides an indication of the reliability and the risk of potential failure in communication with a peer. To the best of our knowledge, there is currently no solution that makes it possible to simultaneously respond to both these problems. In order to resolve this problem, we present a new security model and its protocol based on *blockchain* technology that aims to guarantee not only the validity and integrity of the encrypted authentication data but also aims to associate a scalable and evolutive trust level with the nodes for the lifetime of a network. This model is called the *Blockchain Authentication and Trust Module* (BATM) and allows us to jointly take up authentication and trust management problems in sensor networks. It was initially designed with a view to being deployed on a platform of smart Wisenet cameras (Dubois *et al.* 2016).

This chapter is organized in the following order:

– in section 6.1 we will introduce the IoT challenges that we are going to examine here. These are challenges that are related to functioning, confidentiality and security. We will describe the concepts of authentication and integrity as well as the concepts of reputation and trust. We will then look at the current limitations of these concepts and the factors that must be taken into account when defining our model. We will finally introduce the concept of blockchain, its principal features and how it works as well as how it can be used in a model dedicated to the IoT in order to respond to authentication and trust challenges;

– in section 6.2, we will introduce our contribution, the BATM architecture. We will present the notations used, the general architecture of the BATM and we will describe how BATM aims to respond to authentication needs by specifying the mechanisms that we have implemented. Finally, we conclude the section by introducing the trust and reputation models that we propose to use, as well as the limitations of these models;

– section 6.3 explores the evaluation of BATM architecture through simulations. We present a simulation plan that involves sensor networks operating under BATM, and then interpret the results obtained. Finally, we will provide details regarding how this solution is performed;

– section 6.4 concludes this chapter and discusses the relevance of BATM with respect to the results obtained and also explains the possible future prospects of this work.

6.1.1. *Challenges and motivations, the state of the art*

Data confidentiality and privacy protection are major challenges in IoT networks as the data shared is directly linked to an individual's personal life or to professional secrets for organizations. When these networks are hurriedly set up, it is often at the cost of security. According to HP, up to 70% of these devices are vulnerable to various security breaches (HP 2014). Initiatives such as Linux.Wifatch have demonstrated that the majority of connected objects are susceptible to intrusions by non-authorized third parties, given the absence of any innovative or easy-to-deploy solution.

A large number of recent research projects, or ongoing research, focuses on the security of wireless sensor networks (WSN), especially in the IoT context. There is particular focus on defining the pre-requisites for and constraints around sensor networks in the context of the IoT (Gubbi *et al.* 2013; Sicari *et al.* 2015). Consequently, IoT security is a major challenge for the coming years (Atzori *et al.* 2010; Stankovic 2014; Shah and Yaqoob 2016). This aspect is confirmed by standardization bodies such as IETF and IEEE, which have published a large number of standards dedicated to these networks over the past 5 years. These include 6LowPAN, RPL, CoAP and the DTLS standards, which aim at broadly integrating data authentication, integrity and confidentiality needs as well as mechanisms for availability, resistance to failure and non-repudiation of data, as is the case in the web environment (Granjal *et al.* 2015).

We believe that IoT security must be provided not by approaching each scientific need in an isolated manner but by simultaneously taking into account all the considerations related to the authentication of nodes as well as to trust management. We first present a rapid overview of existing solutions and research on both these aspects as well as the considerations to take into account when defining our model.

6.1.1.1. *Identification and authentication*

Authenticated identification[1] of third parties is an essential component that makes it possible to build trust models, ensuring the non-repudiation of exchanges in the network and making it possible to associate roles and capacities with actors. The objective of authentication is to determine, with certainty, the digital identity of

1 Called "Authentication" in the rest of this chapter, unless indicated otherwise.

adverse parties to allow access control verifications. We will differentiate between existing solutions based on two criteria: (1) the network layer involved and (2) the type of authentication, which is single when only one of the two parties must be authenticated and must be mutual when all parties present authenticate themselves reciprocally. We can cite here the example of a wired IEEE 802.3 network with IEEE 802.1X, which comes into play at the level of the network layer and may be mutually or singly authenticated based on the configuration that is used. Another example is that of authentication during payment on an online commerce site, which comes into play at the application layer level and is mutual, involving participation of a trusted third party.

Authentication over networks is strongly linked to the primitive encryption used to generally authenticate information based on one of the following mechanisms: (1) through the proof of possession of a secret key, in symmetric encryption, or (2) through the proof of possession of a private key in the case of asymmetric encryption. These two solutions are the most widely used ones due to their ability to provide a very high level of proof and as they can be combined in hybrid encryption. In addition to responding to authentication needs in the IoT, these mechanisms can also provide end-to-end encryption, making it possible to simultaneously satisfy the requirements for data confidentiality, integrity and authenticity. Further, the addition of data on identity and date, authenticated in the encrypted flow, make it possible to ensure the non-repudiation of the data. The choice of technology used depends on the objectives of the solution and any constraints that may be present. We will review some properties of each of these methods as follows:

– *Authentication based on symmetric encryption*: this is the most commonly used technology in sensor networks. The joint development of communication standards such as IEEE 802.15.4 (Zigbee) and crypto-processors embedded within nodes have resulted in various advanced encryption standard (AES)-based encryption modes being more widely used (Granjal *et al.* 2015). There is the notable example of the CounTeR (CTR) counter mode, which can guarantee data confidentiality, and the Cipher Block Chaining (CBC) mode, which can authenticate data by generating a Message Authentification Code (MAC) address. The Counter with CBC-MAC (CCM) mode is a combination of the CTR and CBC mode and makes it possible to encrypt as well as authenticate data simultaneously. Depending on the platform, it is possible to use 128-bit, 192-bit or 256-bit keys, although certain norms restrict key size to 128-bits. These modes use a *nonce* or an initialization vector, which must be carefully chosen and must be different for each new encryption using the same key. If a 128-bit key is used, the same nonce must be renewed even for different keys. These constraints require a great deal of rigor in implementing encryption technologies and communication protocols and if this rigor is not applied it could compromise their security. AES-CCM is used in the IEEE 802.15.4 and IEEE 802.11 norms. The chief drawback of using symmetric encryption is the criticality of the key management. The keys must remain secret in order to guarantee the system's properties (integrity, confidentiality and authentication) and at the same time they must be known to both parties. A large number of research projects have approached this problem from two aspects: first, the systems for predistributing keys that allow nodes to possess the keys required for the system to operate. Second, the mechanisms for generating temporary keys or session keys, which are unique for each communication and which enable the creation of private communication channels between two nodes in a network.

– *Authentication based on asymmetric encryption*: using public-key encryption facilitates the generation of a virtual identity associated with a public key via a hashing and signature system that makes it possible to easily verify the information, even offline. This encryption, thus, offers definite advantages, while making it easier to manage the identity of nodes and data authentication (for example, in operations for generating, revoking or updating keys and signatures). Until recently, engineers would generally avoid using solutions based on RSA standards in sensor networks, as the algorithms that were used were too heavy for the embedded nodes. Symmetric algorithms were preferred. These algorithms were part of the standards and it was thus easier to control their implementation for sensors. However, the growing use of public-key encryption algorithms using elliptical curves (ECC) and the fact that these algorithms consumed fewer resources in terms of computing power, memory and energy (Wander *et al.* 2005) made it viable to use them over 8-bit targets and, therefore, in WSNs (Gura *et al.* 2004). This advantage was confirmed by the arrival of a new generation of sensors that were equipped with crypto processors that implemented algorithms based on ECC, which reinforced their viability and led to their being widely adopted. Data authentication and integrity of data were then guaranteed by an electronic signature (ECDSA) and data confidentiality was provided through encryption.

– *Authentication based on hybrid encryption*: the simplification that resulted from public-key encryption algorithms for managing identities and authentication led to new hybrid solutions that used asymmetric algorithms for the first phase of authentication, followed by a step where symmetric keys were generated for securing communications between two nodes. This solution made it possible to combine the advantages of both techniques while cutting down on the disadvantages. These solutions must be particularly resistant in the schemas in which they are used in order to prevent the leaking of any authentication information at each step.

It must be noted that regardless of which technology is considered, it is essential to guarantee the non-divulging of information used for authentication and also to guarantee a key-distribution mechanism that fulfills the confidentiality needs. Existing solutions make use of mechanisms that generate session keys, which are unique and renewable for encrypting information transmitted after the authentication phase is complete. Literature has also provided different methods to implement these features within WSNs (Zhang and Varadharajan 2010; Granjal *et al.* 2015).

6.1.1.2. *Reputation and trust*

The identification of nodes does not prejudge their ability to participate in the smooth functioning of the network and to reply with coherent data that are adapted to the requests that were made. This part is carried out by the trust and reputation management mechanism. The authentication solutions presented in section 6.1.1.1, for instance, largely use asymmetric encryption technologies and implement a verification chain from one or more Certificate Authorities (CA), intrinsically considered to be trust providers. While the chain of digital certificates can authenticate the identity of the participants, it has the disadvantage of being highly centralized, with as many potential major failure points as there are CAs, as each CA must protect itself against any physical or virtual attack that tries to steal their private keys.

PGP adopts a different logic, aiming to authenticate any third party that intervenes on the networks in a decentralized manner using a *Web Of Trust*. This approach, however, cannot be automated in an autonomic network. Consequently, new decentralized topologies for autonomic IoT networks with significant constraints require the definition of new authentication and trust management methods that are autonomic and decentralized.

The contributions presented here show that the authentication of nodes in a network, data integrity, data confidentiality and the non-repudiation of data may be addressed through a crypto-system that is adapted to these needs. However, the reliability of the nodes must be evaluated using a reputation and trust management system dedicated to this task. The characteristics of the architecture and the limitations of the trust management systems for WSNs and the IoT have been discussed in (Nitti *et al.* 2012; Guo and Chen 2015; Chen *et al.* 2016; Fang *et al.* 2016). Reputation systems are a field of active research. Applications such as e-commerce or peer-to-peer networks were initially responsible for propositions such as the beta reputation system proposed in (Josang and Ismail 2002). Trust and reputation management within the IoT are especially important for the following reasons:

– multiple nodes may be providers of one type of data and the quality of data will differ depending on the reliability of the provider that is selected. It is crucial to be able to select the provider that is best suited to the criticality of the application. In order to do this, the reliability of the nodes in the network must be evaluated (Guo and Chen 2015);

– some actors may add nodes whose behaviors could influence the choice of data providers and/or discriminate between third-party nodes in order to exclude certain actors from the network or to consolidate their position (Guo and Chen 2015);

– evaluating trust makes it possible to mitigate internal attacks on the wireless sensor network. It is easy for an attacker to capture a node and study its functioning in order to infiltrate the network. In this case, the harmful behavior of the malicious node must be detected and the use of this node be forbidden (Fang *et al.* 2016).

In addition, trust management systems are targeted by specific attacks, such as the mutual reinforcement of malicious nodes (ballot-stuffing) and false reporting, in order to downgrade the reputation of a node (bad-mouthing). From the preliminary analysis carried out on authentication and trust, we identified three main points to focus on:

1) symmetric encryption seems economically preferable, however the predistribution of keys remains problematic. From this observation, it seems that the hybrid approaches are the most promising;

2) in terms of implementation, it seems optimal to use proven primitives that are widely available and easily interoperable with existing solutions in order to reduce the complexity of the middleware that we propose;

3) existing technologies already allow us to supply authentication mechanisms and guarantee encryption of communication, the confidentiality and authenticity of transfers as well as the trust value that can be attributed to the nodes.

However, existing approaches are confronted with the decentralization of data, which is the chief problem in extending the network. Thus, the problem is not necessarily focused on implementing technological primitives, but rather in organizing access and the verifiability of the decentralized data: implementing an architecture that responds to the challenges that we have discussed requires using a decentralized data structure that is traceable, non-repudiable and which can be authenticated. In this sense, blockchain appears to us to be the most promising data structure to respond to these requirements. As this mechanism is at the heart of our architecture, we will describe the principal characteristics of its functioning in detail in section 6.1.2.

6.1.2. *Blockchain, a support for authentication and trust*

Blockchain is a data structure that was initially designed as a storage support within solutions dedicated to managing decentralized transactions such as Bitcoin (Nakamoto 2008). It is often presented as an accounts book or a decentralized database, but we have generalized its representation here in the form of a logbook that is shared between all nodes in a network. A transaction is defined here as any modification or event that is logged on the blockchain, whether this is a financial application or not. Events are logged in a discrete time model within blocks of valid transactions that have been performed since the last block added into the blockchain. Each block, identified with a number and timestamp, contains new valid transactions, proof of their validity and a MAC that makes it possible to protect its integrity. The blockchain itself is simply the linear chain of these blocks. Only those transactions that are present within a block that is integrated into the blockchain are considered to be definitively valid. The functioning of a blockchain is, thus, based on three distinct and complementary aspects: (1) a network protocol that transmits information between peers; (2) a transactional protocol that is dedicated to managing transactions, including their transmission, verification and their inclusion in a block; (3) a consensus protocol that can ensure a convergence in the system in case of contradictory information between network peers and can determine the current viable state of the chain of blocks. The blockchain is said to be *trustless*, i.e. it minimizes the quantity of trust required by a single actor in the system, by distributing trust between the different actors in the system through an economic game that incentivizes these actors to cooperate with the rules defined by the protocol.

The consensus protocol is a determining feature in the blockchain, which must be given special attention: it makes it possible to guarantee the authentication, non-repudiation and the integrity of the transactions stored on the blockchain and can also solve the problem of double spending. This consensus is made up of a set of rules that all nodes in the network must verify, thereby allowing them to validate the set of events that have taken place and the present state of the chain of transactions. Consensus protocols use two mechanisms to ensure the viability of the transactions. The first is authenticating their viability through a cryptographic signature that provides proof of the ability of a peer to transmit a transaction at a given moment. To date, the majority of blockchain applications use this system to verify the viability of a transaction. The actors on the network are identified by a pair of asymmetric encryption keys, making it possible to define their address, which is a textual representation of the public key in the pair. The private key is used when transmitting a transaction in order to create a signature that can verify the viability of the transaction and guarantee its non-repudiation. The second mechanism consists of guaranteeing the integrity of the logbook. This is done through a *system of proof* whose role is ensuring the continuity of information, preventing double spending, ensuring the integrity of the chain of blocks and preventing byzantine faults. The various existing blockchain technologies use different systems of proof, each with its own characteristics and limitations. The three most commonly used systems are as follows:

1) PoW (Proof of Work): this is the oldest system of proof, the most proven and also the most widespread. It is notably used in Bitcoin. It is based on solving a very complex mathematical problem for which the proposed solution is easy to verify. Although this is the surest system of proof to date, it does present a large number of disadvantages: primarily, the limit on the number of new blocks per hour restricts the speed of validating transactions and prevents it being scaled-up. Further, it is necessary that one part of its computing power be dedicated to solving the mathematical problem, which affects available resources. Increasing the number of miners, with the aim of collecting the fees linked to the creation of a new block and the transaction fees demonstrates the limitations of this system. In April 2018, the Bitcoin network consumed 2.55 GW of energy – as much energy as was consumed by the country of Ireland (de Vries 2018);

2) PoS (Proof of Stake): unlike PoW, PoS does not use any mathematical proof and does not set up a competition between various actors to create new blocks. On

the contrary, it promotes cooperation between actors. A combination of the value possessed the seniority in the network, and a random variable uniquely determines the next creator of a block and a set of validators, who are responsible for verifying the viability of this block and of signing it digitally, if required. There are currently two implementations that use a PoS system considered to be mature: Tezos and EOS, whose protocols are described in their respective technical documentations (Goodman 2014; Eos.io 2018). The Ethereum blockchain is transitioning to a PoS system via the Casper consensus protocol (Zamfir 2017), with the aim of reducing its energy consumption and strengthening the security of the network in the long term. Although it appears promising, this system of proof is often considered to be experimental and is subject to problems that raise questions about its viability (Goodman 2014; Zamfir 2017; Eos.io 2018);

3) PoC (Proof of Correctness): this algorithm specifies, for each server in charge of verifying transactions, a list of third-party servers with which the consensus on the state of the chain is determined. This mechanism gradually validates a state for the blockchain over the entire network. The *Ripple* blockchain uses the PoC system in its Ripple Protocol Consensus Algorithm (RPCA). The chief characteristics of the RPCA are described in its technical documentation (Schwartz *et al.* 2014).

Blockchain has recently aroused a great deal of interest due to its ability to integrate smart contracts. *Smart contract* was initially used to denote an automated contract between different actors that guaranteed that a transaction would proceed smoothly. The generalization of these contracts to implementing decentralized applications within blockchain now make it possible to run code in a verifiable and non-falsifiable manner on the network. This specificity is of particular interest in decentralized networks as it allows the implementation of authentication and trust management systems directly on the blockchain in a *trustless* manner: since the code cannot be modified, the result formally testifies to the present state of the system.

While it is possible, today, to use an entirely standardized protocol stack (Figure 6.1), to the best of our knowledge there is no global approach to security that can be adapted to the different technologies used and to their diverse characteristics. We propose an architecture that is based on blockchain and addresses the following problems:

– authenticating nodes in the network;

– guaranteeing the integrity, confidentiality and non-repudiation of data;

– providing an authenticated encryption solution and an end-to-end encryption mechanism that can create groups of identities and secure access to multicasting services on the network, such as the broadcasting of video flows;

– providing a trust and reputation management system that is resistant to attacks, adaptive and can be deployed as a smart contract in the blockchain.

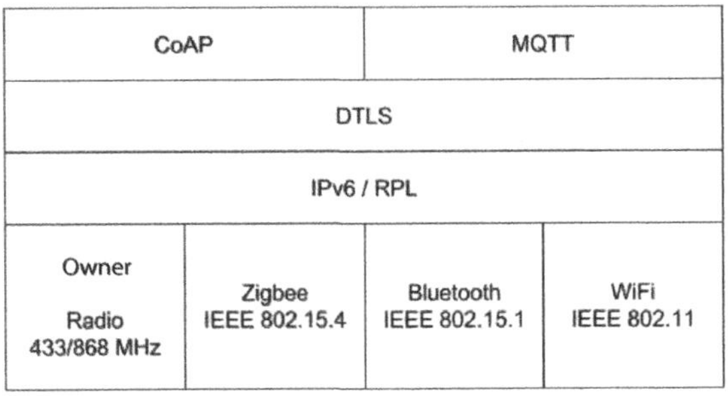

Figure 6.1. *Example of the protocol stack for the IoT*

We called this architecture the Blockchain Authentication and Trust Module (BATM) and it is presented in the next section.

6.2. The Blockchain Authentication and Trust Module (BATM) architecture

6.2.1. *Context and development*

BATM architecture was proposed as a component of the Network Service Execution (NSE) framework, which was developed as part of the Wisenet project (Dubois *et al.* 2016; Marroquin *et al.* 2016). Wisenet is a network of autonomous, smart cameras – a Vision Sensor Network (VSN) – guided by an ontology and used in smart buildings. The NSE has Service-Oriented Architecture (SOA) used to access and redistribute algorithms for processing images (constituting the network services) and whose role is to extract semantic information from the image for contextual decision making. For instance, the network may have several cameras filming the same room in a building and with the following services: detecting people, detecting falls, remotely monitoring vital signals using the remote photoplethysmography (rPPG) method and contacting emergency services. The system is, therefore, able to dynamically organize services to detect the entry of a person into a room. The fall-detection algorithm then transmits information in case of a fall, while the system for remotely monitoring the patient's vitals can monitor the victim's condition and contact emergency services if required. Wisenet is chiefly made up of embedded Raspberry Pi and NVIDIA Jetson TX1 nodes. NSE was then developed for targets with similar computational power and with the goal of providing secure access to services through end-to-end authenticated encryption, secure execution of services, the identification of nodes, reputation and trust management in the system and, finally, the composition of services based on the trust levels of the nodes. In NSE, BATM is the component that is given the responsibility of identification and trust management based on the use of blockchain as a logbook. BATM focuses on carrying out the following tasks:

– logging and updating the cryptographic hardware to identify the nodes;

– tracking node behavior used in calculating trust and reputation;

– developing the protocol and managing system parameters using a voting system (the description of this system is beyond the scope of this chapter);

– the decentralized and non-alterable implementation of algorithms for each of the tasks through the use of smart contracts.

In the next part of this section, we will introduce the notations used and then briefly present the BATM authentication system. We will then focus on our contribution to trust management and on the maximum likelihood trust evaluation (MLTE) algorithm used to evaluate this. Finally, we will present the global architecture of BATM and the interaction between different smart contracts.

6.2.2. *Managing identities and authentication*

BATM makes it possible to identify three types of entities, called the *primaries* in the network: nodes, users and services. Other entities, called *secondaries*, that are registered on the blockchain make it possible to specify the role of each network entity more precisely. Each entity has a digital keychain allowing it to interact over the network. Using a PGP-inspired model, a keychain is made up of a pair of master keys used to authenticate the entity with the BATM, while the three pairs of secondary keys (used, respectively, for signature, encryption and authentication with third-party services) are used in exchanges between pairs. The subkeys are related to the master key through reciprocal signature and we use a system of renewal and revocation of secondary keys similar to the PGP (Callas *et al.* 2007). These keys have a fixed period of validity that is part of the BATM parameters. In case the master key is revoked, it becomes impossible to access the identity and a new identity must be generated.

A BATM identity (Figure 6.2) is a structure composed of a header and a body. The fingerprint of the public master key associated with the identity is used both as a unique identifier (UUID) as well as the proof of possession of the private key. This fingerprint is calculated through the double application of the *sha256* function on the public master key. In order to reduce the risk of Sybil attacks, we use a PoW to add new identities. The *sha256* double hash of the identity structure must satisfy a difficulty parameter (Nakamoto 2008), without which it will be rejected when it is registered, as this system was developed to prevent an attacker with computing power that is greater than the network nodes from registering a large number of false identities on the network.

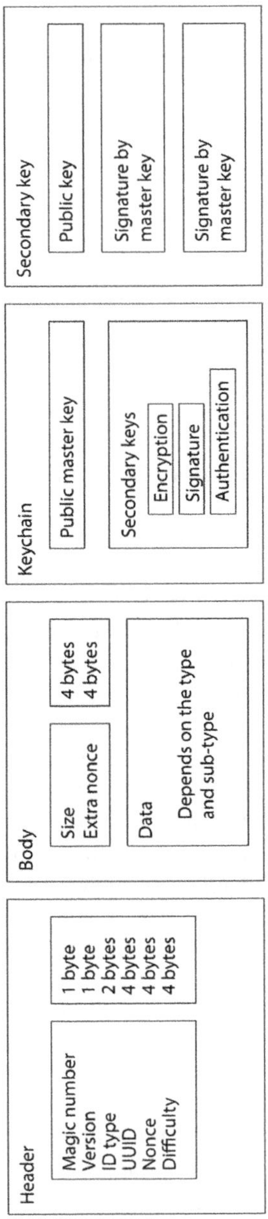

Figure 6.2. *Identity structure and secondary key*

This task requires high computing power and it is very damaging for a node to be required to re-generate an identity. It is recommended here that identities be generated on a machine that has large computational resources and then be transferred to the node. This arrangement may seem limiting, but it can incentivize users and constructors to be more attentive to security. The mechanisms related to users and services are beyond the scope of this chapter and we will therefore focus only on one primary entity – the node.

6.2.2.1. *Registering and managing keys and authentication*

A predeployed smart contract, called an ID Issuer, is responsible for registering a new identity on the blockchain. The parameters in this smart contract are as follows: the new identity structure, the key chain with the public key attached and a signature for each of the structures from the master private key. The validity of a new identity is verified and a new smart contract is published for the new identity.

The structure of this smart contract depends on the primary identity type and its role is to manage the updating of information related to the identity and its cryptographic keys, such as the renewal or revocation of the keys. It is also in charge of distributing the valid secondary public keys to third-party entities for authenticating communications. The ID Issuer smart contract then returns the address of the smart contract generated for the identity. Since this address is the only means of accessing information on the identity in the blockchain, it is essential to store it securely.

When a node initiates an exchange with a peer, it carries out a pairing demand. Figure 6.3 describes the peering process, which must result in the mutual authentication of the nodes.

Exchanging the UUIDs and the addresses for the identity smart contracts for each node makes it possible to recover information to generate the session key and to generate an authentication token characterizing the exchange between two peers. The token is made up of its issue date, for which the hash is signed by both the peers. During the peering, each peer emits a *nonce* value. The hash for each nonce are encrypted using the session key by the remote peer and added to the token. A single token may be used between two peers at any given moment. A new peering request automatically revokes any existing token between two peers. The generated token is valid for requests in both directions in the communication. The session key generated and the token allow the authenticated and non-repudiable encryption of the communication.

Using this, it is possible to create groups of entities, for example to securely access a multicast video flow. In order to do this, we use a multiparty variant of the ECDH algorithm (Giruka *et al.* 2006). For each addition or deletion of an entity in a group, the token and session key are updated to prevent a node being able to decrypt communications that took place before it was added or after it left. The information sent to the group can be decrypted by all nodes that are members of the group.

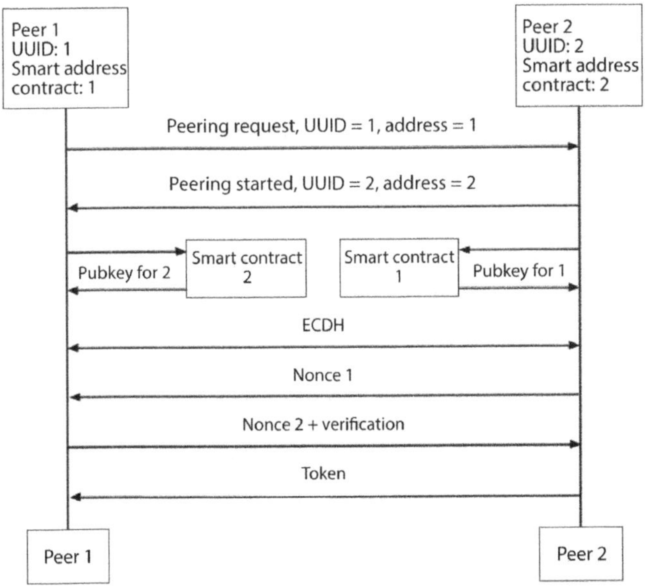

Figure 6.3. *Peering process*

6.2.3. *Calculating trust and reputation using the MLTE algorithm*

In the BATM, trust and reputation are only applicable to node-type entities. BATM provides both the computing engine for reputation and trust in the network as well as the mechanism to propagate trust values (Josang and Ismail 2002). Using the classification system proposed in (Guo and Chen 2015), we list the characteristic principles of BATM as a trust management system as follows:

– the composition of reputation and trust values is based on the quality of service of a node, defined as its propensity to provide usable data, and the user may partly influence the final calculation of trust;

– the propagation of trust is decentralized; in our case, it is implemented by smart contracts in the blockchain. The aggregation of input data are carried out through Bayesian inference, followed by a weighted mean and updating through a linear combination of fixed parameters;

– the updating of the system takes place in discretized periods of time through the collection of information on interactions that took place since the last update. The nodes requesting a service give feedback on their respective providers.

The goal of our model was to respond to all the trust problems that were described in the Wisenet project, as well as the threat model defined in (Chen *et al.* 2016). We will start by defining the concepts of trust and reputation within the network based on the work produced by Gambetta *et al.* (2000): trust and reputation are modeled here by two estimations of a probability that estimates the accuracy of a future action. *Reputation* represents the consensual probability (identical throughout the network) of the accuracy of an action being carried out by a third-party node. As a sensor is not able to provide more than one dataset, usable or unusable, we limit the outcome of an action to "success" or "failure". This limitation does not allow us to represent various errors and only takes into consideration the result of an action, which we judge to be sufficient to characterize the reliability of a node. Each of these actions is compared to a random variable in the set $0, 1$, the result of a Bernouilli test, where the outcome is characterized by success, 1, or failure, 0. This basic hypothesis is similar to that chosen by (Josang and Ismail 2002; Chen *et al.* 2016), who only considered binary events. Unlike (Josang and Ismail 2002; Chen *et al.* 2016), we use an estimator of maximum likelihood, denoted by ML, and not a beta distribution for probability density. The algorithm presented here is a complete and enhanced version of the preliminary study discussed in (Moinet *et al.* 2017, 2018).

In order to calculate this estimator, ML, we propose a new algorithm in the next section, which we call the MLTE, whose goal is to calculate the reputation and trust of a node. We will present the notations used in the MLTE algorithm and then present an overview of the algorithm itself, before describing each step used in the calculation. We will then describe its implementation in the form of smart contracts.

6.2.3.1. *Notations*

The notations used in the MLTE algorithm are: $N = \{n_1, n_2, \dots, n_K\}$, which denotes the set of nodes in the network with cardinality K. Each node is characterized with respect to its peers using a distance metric lying in $[0, 1]$. We use the Euclidian distance $dist(n_i, n_j)$ between two nodes n_i and n_j as metric. The spatial localization of any node n_i using its spatial coordination, in the form of a triplet (x_i, y_i, z_i), is considered to be available *a priori* for calculating this metric. Our research does not concern the location of nodes. We thus recommend that the reader refer to available work in this field, such as (Mao *et al.* 2007; Deak *et al.* 2012; Chowdhury *et al.* 2016; Chelouah *et al.* 2017).

The MLTE algorithm operates through the discretization of time. We thus consider t to be the time variable, as a sample $t \in \mathbb{N}^+$.

Each node n_i has its own initialization time corresponding to its time of registration $t = T_{Ej}$. For simplicity's sake, we will assume in the rest of this section that all nodes are registered at $t = 0$. T denotes the final sampling carried out, corresponding to the final execution of BATM. The sampling period for the system characterizes the frequency of execution of MLTE. It is important to distinguish between the *sampling frequency*, t, of the system and the *calculation frequency* between two sampling periods, t_1 and t_2.

An action, denoted by $a_{t,c}^{n_i \to n_j}$, represents an interaction that occurred over a period t between the node n_i that requested a service and the node n_j that provided the service and this is numbered in its period using the index $c \in \mathbb{N}^{+*}$.

This index discriminates between different actions that take place in the same time period. We use $A_t^{n_i \to n_j} = \{a_{t,0}^{n_i \to n_j}, \ldots, a_{t,C_t^{n_i \to n_j}}^{n_i \to n_j}\}$ to denote the set of actions between the requester node n_i and the provider node n_j over the period t, with $|A_t^{n_i \to n_j}| = C_t^{n_i \to n_j}$ denoting this number of actions. An event, denoted by $e_{t,c}^{n_i \to n_j} \in \{0, 1\}$, characterizes the outcome of the action $a_{t,c}^{n_i \to n_j}$, and takes the value 1 for a success or the value 0 for a failure. Similar to the system used for actions, we use $E_t^{n_i \to n_j} = \{e_{t,0}^{n_i \to n_j}, \ldots, e_{t,C_t^{n_i \to n_j}}^{n_i \to n_j}\}$ to denote the events resulting from actions between n_i and n_j over a period t.

The number of successes between n_i and n_j over a sampling period t is denoted by $S_t^{n_i \to n_j}$:

$$S_t^{n_i \to n_j} = \sum_{c=0}^{C_t^{n_i \to n_j}} e_{t,c}^{n_i \to n_j} \qquad [6.1]$$

The number of events and the number of successes over a period stretching between t_1 and t_2 are, respectively, denoted by $NE(n_i \to n_j, t_1, t_2)$ and $NS(n_i \to n_j, t_1, t_2)$:

$$NE(n_i \to n_j, t_1, t_2) = \sum_{t=t_1}^{t_2} C_t^{n_i \to n_j} \qquad [6.2]$$

$$NS(n_i \to n_j, t_1, t_2) = \sum_{t=t_1}^{t_2} S_t^{n_i \to n_j} \qquad [6.3]$$

The overall number of events between n_i and n_j is denoted by $NEG(n_i \to n_j, t_1, t_2)$ and this is defined for all nodes n_j for which $NE(n_i \to n_j, t_1, t_2) \neq 0$:

$$NEG(n_i \to n_j, t_1, t_2) = NE(n_i \to n_j, t_1, t_2) + NE(n_j \to n_i, t_1, t_2) \qquad [6.4]$$

Finally, the total number of events that occurred at the node n_i between the periods t_1 and t_2, denoted by $NAE(n_i, t_1, t_2)$, is defined by:

$$NAE(n_i, t_1, t_2) = \sum_{j=0}^{K} NEG(n_i \to n_j, t_1, t_2) \qquad [6.5]$$

The configuration of the execution of MLTE for all the nodes takes place across six global parameters, presented in Table 6.1.

Parameter	Role
τ	The basic reputation
η	Roll-of factor
α	Weighting factor for local success
β	Weighting factor for importance
Γ	Metric of maximum distance
S	Size of the calculation windows

Table 6.1. *Global parameters for BATM*

6.2.3.2. *The general functioning of the MLTE algorithm*

The MLTE algorithm can be divided into three sequential steps. Each of these steps is executed on all nodes, n_i, in the network.

– *Step 1, evaluating peers*: each requester node estimates the probability of success of each of its providers. Using the values for events, the requester nodes calculate an estimator for the (Maximum Likelihood) $ML(n_i \rightarrow n_j, t_1, t_2)$ of the provider's probability of success. However, this rough estimation does not allow the behavior of the node to be represented, does not provide a reference value and does not decrease in case of inactivity. $ML(n_i \rightarrow n_j, t_1, t_2)$ is, thus, used to calculate the reliability index $I(n_i \rightarrow n_j, t_1, t_2)$, characterizing the opinion of the requester node, i, with respect to a provider, j and yielding a coherent value of the behavior of j as seen by i. The estimation cannot be carried out in the absence of events and thus $ML(n_i \rightarrow n_j, t_1, t_2)$ and $I(n_i \rightarrow n_j, t_1, t_2)$ are defined for every pair of nodes i, j such that $NE(n_i \rightarrow n_j, t_1, t_2) \neq 0$.

– *Step 2, evaluating its own reputation*: each node calculates its own reputation based on the reliability indices calculated by all its peers. For each node, i that has requested a service from j, j recovers the reliability index $I(n_i \rightarrow n_j, t_1, t_2)$ that it weights based on the level of local success of i and the level of importance of i in all its exchanges. The mean of its weighted indices make up the reputation of the node j, denoted by $R(j, t_1, t_2)$, in the network.

– *Step 3, calculating the trust level toward peers*: each of the nodes, whether it is a requester, provider or passive, calculates the trust level that it attributes to each of its peers in the network. A node i calculates the trust level toward j, denoted by $\Phi(n_i \rightarrow n_j, t_1, t_2)$, using the reputation of j, $R(j, t_1, t_2)$, the distance metric between i and j and a user factor δ.

Each of these steps is presented in detail in the next part. It must be noted that all the results obtained from MLTE are values lying between $[0, 1]$. We use a windowing method for the events in order to avoid too much inertia in the values calculated over the period between t_1 and t_2. As the calculations are based on the sums of the events, the greater the number of events considered, the lower the importance of recent events considered. This harms the reactivity of the system and increases its inertia. To remedy this problem, our approach uses three rotating windows of size S, being a multiple of 3, for each node. S must be identical for all nodes in the network as the number of accumulated events is used in the algorithm. They are initialized in cascade at $t = 0$, $t = S/3$ and $t = (2S)/3$, and reinitialized once they are filled. We use three states for each window:

1) *initialization*: at $t = t_1 + S$, we fix $t_1 = T$. All values in the last calculation period are reinitialized. The window once again begins to accumulate events from t_1 onwards;

2) *refining*: at $t = t_1 + S/3$, the number of samples still does not allow pertinent estimation of the values using MLTE and the window continues to accumulate events;

3) *use*: at $t = t_1 + (2S)/3$, the window has accumulated a sufficient number of events to accurately estimate the node's reputation and trust.

Upon the initiation of MLTE, the values for each node must be considered as unreliable until $t = 2S/3$, when the first window is able to provide an accurate estimate of values. The window in use at the moment is then reinitialized and the next window becomes a reference every $S/3$ units of time in order to benefit from the best compromise between the accuracy of estimation and adequate reactivity. Figure 6.4 illustrates how the windowing operates.

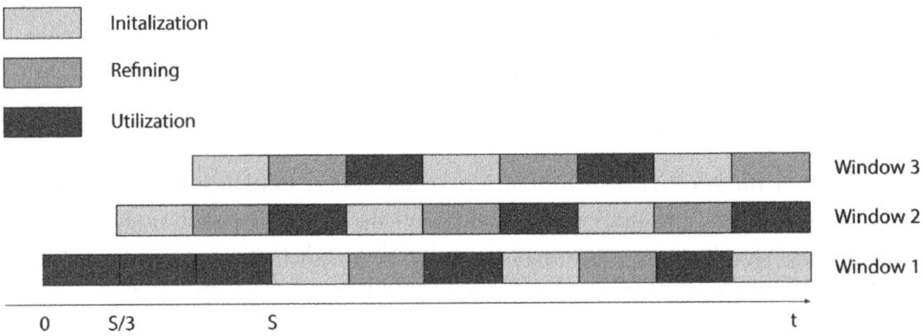

Figure 6.4. *The processing window*

6.2.3.3. *How the MLTE functions: a detailed discussion*

In this section, we will discuss each of the sequential steps of the MLTE and describe the operations carried out at each step, how the different indicators are calculated and the decision-making processes when local events occur.

Step 1: Estimating the maximum likelihood and the reliability index

The maximum likelihood estimator $ML(n_i \to n_j, t_1, t_2)$ gives the probability of the success of the future actions of n_j viewed by n_i at the time t_2, taking into consideration the events that occurred since t_1. Its value is equal to the mean of the successes S with respect to the total number of events NE. We use the iterative form of $ML(n_i \to n_j, t_1, t_2)$ for implementation, as given in equation [6.6], writing $ML(n_i \to n_j, t_1, t_1) = 0$ as per convention:

$$ML(n_i \to n_j, t_1, t_2) = \frac{(ML(n_i \to n_j, t_1, t_2 - 1) \times NE(n_i \to n_j, t_1, t_2 - 1)) + S_{t_2}^{n_i \to n_j}}{NE(n_i \to n_j, t_1, t_2)} \quad [6.6]$$

Calculating the estimation $ML(n_i \to n_j, t_1, T)$ for each node n_i for all pairs $n_j \in N$ for which $NE(n_i \to n_j, t_1, T) \neq 0$ does not make it possible to define a mean reliability level, representing the value toward which the computations converge when the system does not have enough information (Chen *et al.* 20106a). Further, as specified by Nitti *et al.* (2012), it is necessary to prioritize recent information during evaluation in order to enable the Right to Forget in the case of a temporary failure in a node, and also to increase the system's reactivity in the case of a sudden change in behavior. Existing research applies a decreasing factor to past events, usually represented by an exponentially decreasing function. BATM uses a different approach, using a constant denoted by τ defining a base value toward which the computations converge in the absence of events.

The *reliability index* $I(n_i \to n_j, t)$ makes it possible to define a representative value for the behavior of each node. It is expressed as a series whose formulation is governed by the occurrence of an event at time $t = T$, with $I(n_i \to n_j, t_1, t_1) = \tau$. When a new event is detected, the index is computed by weighing the distance from ML to τ by a factor $\frac{1}{1+NE(n_i \to n_j, t_1, t_2)}$, as presented in equation [6.7]:

$$I(n_i \to n_j, t_1, t_2) = \frac{\tau - ML(n_i \to n_j, t_1, t_2)}{1 + NE(n_i \to n_j, t_1, t_2)} + ML(n_i \to n_j, t_1, t_2) \quad [6.7]$$

If no new event occurs ($C_T^{n_i \to n_j} = 0$), the distance between τ and $I(n_i \to n_j, t_1, T - 1)$ is reduced by a factor η to ensure convergence toward τ, as given in equation [6.8]:

$$I(n_i \to n_j, t_1, t_2) = \eta \times \tau + (1 - \eta) \times I(n_i \to n_j, t_1, t_2 - 1) \quad [6.8]$$

It must be noted that in our system, the nodes cannot transmit events (and thus the reliability indicators) for themselves. This prevents attacks through the auto-promotion of trust levels.

Step 2: Calculating the reputation R(j, t_1, t_2)

Each of the nodes j calculates its reputation within the network. Once all the nodes n_i requesting a service during the calculation period have updated their reliability index $I(n_i \rightarrow n_j, t1, t2)$, j proceeds to calculate its reputation $R(n_j, t_1, t_2)$. The reputation of a node n_j is calculated as being the mean of its reliability indices $I(n_i \rightarrow n_j, t1, t2)$, weighted by two indicators, the *local success ratio* SR and the *importance ratio* IR, which are defined below:

– the *local success ratio* $SR(n_j, n_i, t_1, t_2)$ uses the parameter α for weighting the success ratio for past exchanges between n_i, a service requester, and n_j, the service provider. This parameter makes it possible to prioritize the index for nodes n_i for whom the number of past successes, $NS(j, i, t_1, t_2)$, is higher:

$$SR(j, i, t_1, t_2) = \frac{1 + NS(j, i, t_1, t_2)}{1 + NE(j, i, t_1, t_2)} \times \alpha \qquad [6.9]$$

– the *importance ratio* $IR(j, i, t_1, t_2)$ uses the parameter β for weighting the importance of a node in the exchanges carried out. This parameter penalizes nodes that have recurring exchanges. The more preponderant n_i is in interactions of n_j, the more its trust value is attenuated. This ratio is particularly useful against bad mouthing and ballot-stuffing attacks:

$$IR(j, i, t_1, t_2) = \frac{1 + NAE(j, t_1, t_2)}{1 + NAE(j, t_1, t_2) + NEG(j, i, t_1, t_2)} \times \beta \qquad [6.10]$$

The equality $\alpha + \beta = 1$ makes it possible to keep the scale of values to within $[0, 1]$ for calculating the reputation $R(n_j, t_1, t_2)$. This type of weighting makes it possible to give greater importance to nodes whose trust value is higher. This method is widely used in the literature (Shafer 1976; Jøsang 2001) as a means to reinforce the weight of the most reliable nodes.

Let F_{n_j} be the set of nodes with a non-empty index $I(n_i \rightarrow n_j, t_1, t_2)$. The calculation of the reputation $R(n_j, t_1, t_2)$ is defined over the nodes of the set F_{n_j} as given below:

$$R(n_j, t_1, t_2) = \frac{\sum_{i=0}^{i=K} I(n_i \rightarrow n_j, t_1, t_2) \times (SR(j, i, t_1, t_2) + IR(j, i, t_1, t_2))}{|F_j|} \qquad [6.11]$$

When $F_{n_j} = \emptyset$, we impose $R(n_j, t_1, t_2) = \tau$.

As the calculation of $I(n_i \rightarrow n_j, t_1, t_2)$ is correlated with the number of exchanges carried out and the calculation of the reputation using the evaluation over the set of nodes, it becomes more difficult to carry out ballot-stuffing attacks, especially when the nodes have a wide variety of different providers.

Step 3: Calculating trust toward pairs

The reputation value is a global and consensual indication of how reliable a node in the network is. The calculation of a trust value Φ is specific to a pair of nodes and is not reciprocal, i.e. $\Phi(n_i \rightarrow n_j, t_1, t_2) \neq \Phi(n_j \rightarrow n_i, t_1, t_2) | n_i, n_j \in N, n_i \neq \{n_j\}$. The calculation of trust must, therefore, integrate two subjective parameters: Γ and δ, which depend on the node carrying out the calculation and with each playing a specific role, which we describe below:

– the parameter Γ represents the largest possible distance in the network's geographic deployment zone. The distance ratio $RD(n_i, n_j)$ characterizes the distance between two nodes n_i and n_j. Using the distance metric $dist(i, n_i, n_j j)$, this distance must be normalized to 1 using Γ:

$$RD(n_i, n_j) = (1 - ((\frac{dist(n_i, n_j)}{\Gamma}) \mod 2)) \qquad [6.12]$$

– the parameter δ is a user factor that makes it possible to manually promote or penalize exchanges with certain nodes in the network. The possible values for δ are specified in the finite set D, which is part of the BATMparameters.

Calculating trust is a linear combination of these two parameters as shown in equation [6.13]:

$$\Phi(n_i \rightarrow n_j, t_1, t_2) = R(n_i \rightarrow n_j, t) \times RD(n_i \rightarrow n_j, t_1, t_2) \times \delta \qquad [6.13]$$

6.2.3.4. *Selecting the provider based on trust*

The trust value obtained for each of the nodes is used to select the best possible service provider. For each new request, the requesting node n_i determines the set $FS \subset N$ of all potential providers of the requested service. The chosen provider is the provider with the highest trust value $\Phi(n_i \rightarrow n_j, t_1, T)$. Since the distance metric is taken into account when calculating trust, this value guarantees the best possible compromise between quality of service and distance. It must be noted that in the case of BATM, the composition of services is specified by the Wisenet ontology. The set FS, therefore, contains the providers that satisfy the ontology's composition criteria, for instance, the maximum distance allowed.

6.2.3.5. *Integrating MLTE in BATM using* smart contracts

The set of global MLTE parameters presented in Table 6.1 are stored within a smart contract that is predeployed in the blockchain. Each of the three steps presented is implemented by a smart contract. The structure of the smart contracts is given in Figure 6.5. When a new identity is recorded by a node, the ID Issuer contract publishes a contract dedicated to managing log reports and to storing the values used in the calculations for this node.

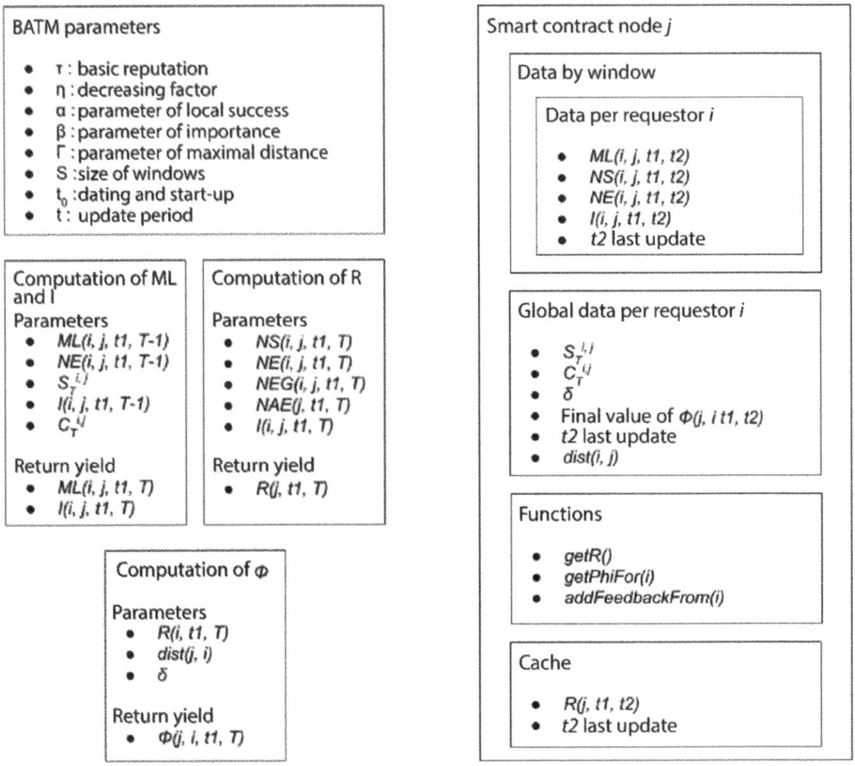

Figure 6.5. *Trust calculation – smart contracts*

The date and time stamping that is carried out when the system starts up allows the new contract to define the time for the next calculation as well as the time for the rotation of the windows in synchronization with the other nodes, so as to guarantee the coherence of calculations in the network. The values are computed on demand, when one of the functions (*getR()* or *getPhiFor(i)*) in the contract for a node is accessed, if $T - t_2 > t$. This makes it possible to reduce the power of the computation used by BATM compared to an execution over each update period, which is potentially meaningless if $R(j, t_1, T)$ or $\Phi(j, i, t_1, T)$ are never accessed. Similarly, the maximum likelihood estimator $ML(n_i \to n_j, t_1, T)$ and the reliability index $I(n_i \to n_j, t_1, T)$ are updated when any feedback is reported via the *addFeedbackFrom(i)* function or during access to the reputation $R(j, t_1, T)$. In order to secure access to the node contract, the functions *getR()*, *getPhiFor(i)* and *addFeedbackFrom(i)* must verify the identity of the caller. A *Chain Token* generated by the contract (the implementation of which is beyond the scope of this chapter) enables this verification.

To conclude this section, the heart of the BATM is made up of six smart contracts that are predeployed over the blockchain and given the responsibility of managing identities and running the MLTE. Each entity has two smart contracts dedicated to managing its identity and its encryption keys. These contracts are also responsible for accessing the trust and reputation values and logging the feedback from the nodes requesting the service.

In order to retrace the execution of the algorithms, the different functions related to the smart contracts for authentication and trust transmit transactions recorded on the blockchain that characterize the event received or the update carried out. These data are not used by the computation algorithms but may be used by any node that wishes to control the smooth functioning of the system.

6.3. Evaluating BATM

In the following section, we will discuss the evaluation of the BATM architecture and the performances offered by the MLTE algorithm over simulations on a sensor network that was subject to the same constraints as the IoT. We chose to evaluate MLTE through simulations using the python SageMath framework (Sage 2018). The goal of this evaluation was to characterize the performance of the algorithm on a simplified network model in order to validate its performance and then to evaluate its resistance to node failure.

We will first introduce our simulation plan and then discuss the results obtained and the interpretation of these results.

6.3.1. *Simulation plan*

We chose nine services, considered independently and distributed at random over a network composed of 30 nodes whose positions were fixed and generated randomly on a 3D geographic zone with the dimensions $30 \times 30 \times 30$ m. To keep things simple, we considered that all nodes could communicate directly with each other and did not integrate aspects of routing into our simulation. However, it is perfectly possible to integrate constraints of this kind and this has no significant influence on the computations of the proposed estimators.

Each node is the provider of a set of services and has a probability of success that is used to draw the events that it supplies. Requests are generated for each update period on the network, with each of the nodes requesting a service. We thus had 30 service requests for every period, t. This choice represents more activity than is generally observed, however it is relevant in order to carry out an initial evaluation of our model as we have a greater number of events available to us. Over every period, t, the selection of services per trust level determined the provider for each requester. For the purposes of this evaluation, it is considered that t is an atomic unit of time. The corresponding events are generated and MLTE is run on the network to update the reputation and trust values. We simulated MLTE over 1,000 updates periods t of the algorithm.

To make it easier to understand the study, the values for $ML(n_i \rightarrow n_j, t_1, T)$ and $i(n_i \rightarrow n_j, t_1, T)$ presented in Table 6.2 are the mean values carried out over the set of requesting nodes for the node observed. The ground truth is the probability of success for the node under consideration. The global parameters were fixed (see Table 6.2) such that they would favor exchanges with multiple providers, hence the choice of β over α. The size of the window was chosen in order to obtain the best possible compromise between the system's stability and reactivity. Since the values for τ and η were not essential for the general evaluation of the algorithms, they were chosen in order to minimize their impact on the observation of the results.

Parameter	Value
τ	0.5
η	0.1
α	0.4
β	0.6
S	300

Table 6.2. *BATM simulation parameters*

6.3.2. *Results and interpretation*

We will first observe the behavior of the system in a simulated sensor network and then evaluate the performance of the MLTE. We will then proceed to describe the behavior of the system in a scenario where there is node failure.

6.3.2.1. *Evaluating MLTE in ideal conditions*

We will evaluate the behavior of the system operating under MLTE in an ideal scenario, with no node failure. Figure 6.6 presents MLTE estimations for a node, j. This represents the average of the estimations $ML(n_i \rightarrow n_j)$ (orange curve), the average of the reliability indices $I(n_i \rightarrow n_j, t_1, t_2)$ (yellow curve) and the reputation $R(n_j, t_1, t_2)$ of the node j (green curve) for MLTE operations during the calculation window. The ground truth (blue curve) is the probability of success of the node.

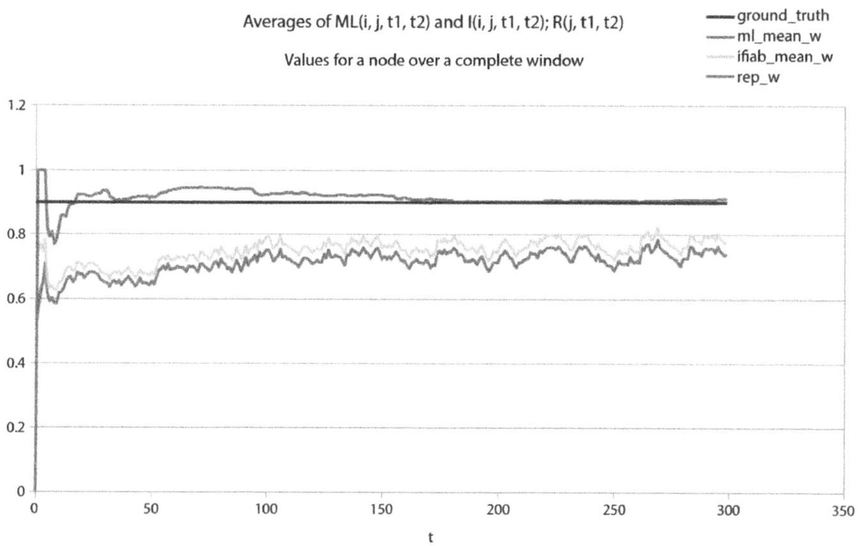

Figure 6.6. *Complete calculation window. For a color version of this figure, see www.iste.co.uk/mbarek/service.zip*

We observe three consecutive phases of the window. In the first 100 samples, the estimations oscillate and are imprecise. The second phase shows the refining of the values between t = 100 and t = 200, with less than 3% error between $ML(n_i \rightarrow n_j)$ and the ground truth at t = 150. During the use phase, the values are stable and the estimation $ML(n_i \rightarrow n_j)$ remains at less than 3% error. This supports our choice in the size of the window and processing for our evaluation. We can also see that the evolution of the mean reliability index for the node as well as its reputation are correlated in the absence of any event over a period t, as well as in the absence of any failures generated by the node, which reinforces the advantages of our system. These initial tests show a mean error in estimating the failure level of 10.3%, with a median of 5%. The mean error for a node varies between 0.3% and 30%. It must be noted that this value corresponds to a node with low activity. With 30 events for 30 nodes in each period, t, an average node must record around 1,000 events on average, over 1,000 sampling periods. This latter node has only recorded 630 times across all the pairs, which explains the larger error level and the gap between the mean and the median for the error levels. The mean reliability index and the reputation value demonstrate the stability of BATM over time.

We have also evaluated the performance of the selection of service providers based on trust (this evaluation is not presented here). To do this, we compared the choice made by MLTE with the optimal node determined through comparing the probability of success among the nodes. It was observed that MLTE selected the optimal node in 76% of cases, which shows that it is reliable and also reactive in selecting the best possible provider. Indeed, each time a node malfunctioned, the probability of this node being selected reduced, even if it had a very high probability of success. On the other hand, a node with low reliability but whose most recent events were successful saw an increase in its probability of being selected. These two factors explain the difference in choosing the optimal node in the remaining 24% of cases.

6.3.2.2. *Behavior of the MLTE in the event of node failure*

The MLTE was also confronted with a scenario with a node failure. The probability of success was fixed at 0.9 when the system started up. Halfway through the simulation time this was modified to 0.2 in order to simulate the failure of a node and observe how the BATM reacted to this node, as shown in Figure 6.7. In this figure, we can observe the same indicators as in the previous figure along with the averages of the $ML(n_i \rightarrow n_j)$ using a non-windowed method (light blue curve). This figure shows the evolution of the indicators $ML(n_i \rightarrow n_j, t_1, t_2)$, $I(n_i \rightarrow n_j, t_1, t_2)$ and $R(n_j, t_1, t_2)$ for the malfunctioning node j. The node j was scheduled to fail at $t = 500$.

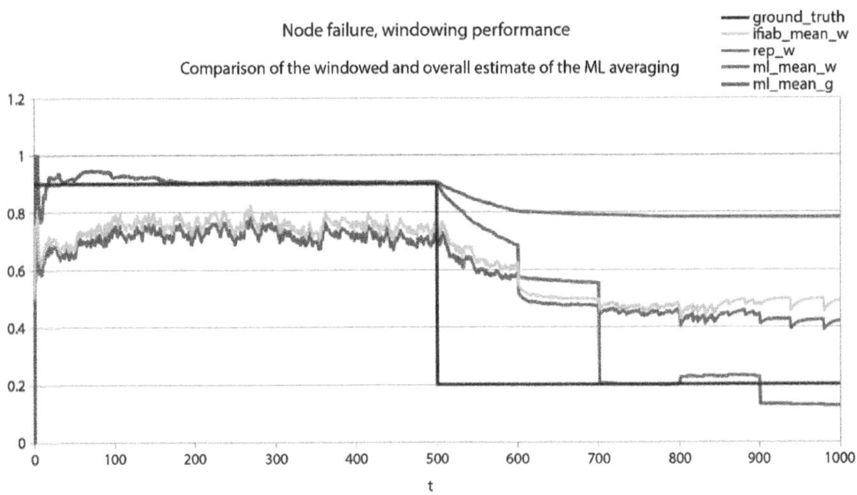

Figure 6.7. *Simulation of failure of a node. For a color version of this figure, see www.iste.co.uk/mbarek/service.zip*

We can note the advantages of using windowing as compared to using a more global method. In the latter, the inertia induced by the events that occurred at the start of the simulation stabilize the estimation and prevent the system from reacting. On the other hand, the windowed system allows the system to react quickly. The fall in the probability level from 77% brings about an average drop of 24% and 38% in the ML

on the first two rotations of the windows and then an alignment on the new probability of the success. The reputation falls from 19% and 34% over the first two rotations of the window. We then observe a larger error in estimating the probability of success and a periodic behavior in the value of the reputation of the node around the value of τ. This behavior is due to a small number of events starting from around $t = 600$. The node with a low reputation is rarely selected by its peers to provide a service. However, the convergence toward the base value for reputation allows the node new chances to re-establish higher values for trust and reputation. These results confirm how reactive BATM and MLTE are to sudden changes in the network.

6.4. Conclusion

This chapter provides the context and existing research on authentication and trust management in decentralized wireless networks. The BATM architecture was proposed as an innovative solution to the problems that were encountered. The results from the reputation and trust model testify to the advantages of our solution and of blockchain in storing events related to the network and the behavior of nodes in the IoT. As far as we are aware, BATM is the only solution that addresses problems in both authentication as well as trust management and that uses blockchain as an information support.

The performance of BATM enables us to move on to the next step of carrying out a proof of concept on one of the leading blockchain technologies supporting smart contracts, such as Tezos, EOS or Ethereum. This implementation will be assessed using a network of Wisenet cameras used within the LE2I laboratory in the University of Bourgogne.

One of our objectives going forward is to be able to authenticate nodes via BATM as soon as the network is accessed. Using group authentication mechanisms, it is possible to create a group corresponding to the local network that carries out this function. Adding an authentication service for IEEE 802.1X based on this principle will make it possible to use it with IEEE 802.3 (Ethernet) and IEEE 802.11 (WiFi) norms. For 802.11s wireless mesh networks, we examined a possibility based on mechanisms used for Simultaneous Authentication of Equals (SAE). Access to the network then requires that a node that requests a connection already possess a BATM identity. This can be provided by out-of-band registration, through radio frequency identification (RFID), for instance. These additional safety mechanisms must be provided to mitigate the possibility of spoofing and to prevent a captive network being created by an attacker.

Similarly, the MLTE algorithm must be evaluated in different conditions. For example, a low network load (thus, a small number of events) may lead to certain parameters being modified. The network's behavior on the addition or deletion of nodes as well as its response to attacks must also be evaluated.

6.5. References

Akyildiz, I.F., Su, W., Sankarasubramaniam, Y., and Cayirci, E. (2002). A survey on sensor networks. *IEEE Commun. Mag.*, 40(8), 102–105.

Atzori, L., Iera, A., and Morabito, G. (2010). The Internet of Things: A survey. *Comput. Netw.*, 54(15), 2787–2805. Available at: http://linkinghub.elsevier.com/retrieve/pii/S1389128610001568.

Callas, J., Donnerhacke, L., Finney, H., Shaw, D. and Thayer, R. (2007). Openpgp message format (rfc 4880). Available at: http://tools.ietf. org/html/rfc4880.

Chelouah, L., Semchedine, F., and Bouallouche-Medjkoune, L. (2017). Localization protocols for mobile wireless sensor networks: A survey. *Comput. Electr. Eng.*, 59, 733–751.

Chen, R., Guo, J., and Bao, F. (2016). Trust management for SOA-based IoT and its application to service composition. *IEEE T SERV COMPUT.*, 9(3), 482–495.

Chowdhury, T.J., Elkin, C., Devabhaktuni, V., Rawat, D.B., and Oluoch, J. (2016). Advances on localization techniques for wireless sensor networks: A survey. *Comput. Netw.*, 110, 284–305.

de Vries, A. (2018). Bitcoin's growing energy problem. *Joule*, 2(5), 801–805.

Deak, G., Curran, K., and Condell, J. (2012). A survey of active and passive indoor localisation systems. *Comput. Commun.*, 35(16), 1939–1954.

Dubois, J., Moinet, A., Bobbia, S., Marroquin, R., Heyrman, B., Bonazza, P., Darties, B., Nicolle, C., Benezeth, Y., Mitéran, J., and Ginhac, D. (2016). WiseEye: A platform to manage and experiment on smart camera networks. *5th Workshop on Architecture of Smart Cameras (WASC)*, Dijon, France.

Eos.io (2018). Eos.io technical white paper v2. Available at: https://github.com/EOSIO/Documentation/blob/master/TechnicalWhitePaper.md#consensus-algorithm-bft-dpos.

Fang, W., Zhang, C., Shi, Z., Zhao, Q., and Shan, L. (2016). Btres: Beta-based trust and reputation evaluation system for wireless sensor networks. *J. Netw. Comput. Appl.*, 59, 88–94.

Gambetta, D. *et al.* (2000). Can we trust trust. *Trust: Making and breaking cooperative relations*, 13, 213–237.

Gartner. (2017). Gartner Says 8.4 Billion Connected "Things" Will Be in Use in 2017, Up 31 Percent From 2016. Available at: https://www.gartner.com/newsroom/id/3598917.

Giruka, V.C., Chakrabarti, S., and Singhal, M. (2006). A distributed multi-party key agreement protocol for dynamic collaborative groups using ecc. *J. Parallel Distrib. Comput.*, 66(7), 959–970.

Goodman, L. (2014) Tezos, a self-amending crypto-ledger white paper. Available at: https://tezos.com/static/papers/white_paper.pdf.

Granjal, J., Monteiro, E., and Silva, J.S. (2015). Security for the internet of things: A survey of existing protocols and open research issues. *IEEE Commun. Surv. Tut.*, 17(3), 1294–1312.

Gubbi, J., Buyya, R., Marusic, S., and Palaniswami, M. (2013). Internet of Things (IoT): A vision, architectural elements, and future directions. *Future Gener. Comp. Sy.*, 29(7), 1645–1660.

Guo, J. and Chen, R. (2015). A classification of trust computation models for service-oriented Internet of Things systems. *2015 IEEE International Conference on Services Computing (SCC)*, IEEE, New York City, NY, USA, 324–331.

Gura, N., Patel, A., Wander, A., Eberle, H., and Shantz, S.C. (2004). Comparing elliptic curve cryptography and rsa on 8-bit cpus. *International workshop on cryptographic hardware and embedded systems*, Springer, Cambridge, MA, USA, 119–132.

HP. (2014). Press release, 29 July 2014. Available at: http://www8.hp.com/us/en/hp-news/press-release.html?id=1744676.

Josang, A. and Ismail, R. (2002). The beta reputation system. *Proceedings of the 15th Bled Electronic Commerce Conference*, Bled, Slovenia, 5, 2502–2511.

Karl, H. and Willig, A. (2005). *Protocols and Architectures for Wireless Sensor Networks*. John Wiley & Sons, Ltd, Chichester, UK. Available at: http://doi.wiley.com/10.1002/0470095121.ch1.

Mao, G., Fidan, B., and Anderson, B.D. (2007). Wireless sensor network localization techniques. *Comput. Netw.*, 51(10), 2529–2553.

Moinet, A., Darties, B., and Baril, J. (2017). Blockchain based trust & authentication for decentralized sensor networks. *CoRR*, abs/1706.01730. Available at: http://arxiv.org/abs/1706.01730.

Moinet, A., Darties, B., and Baril, J. (2018). BATMAN: plate-forme blockchain pour l'authentification et la confiance dans les wsns. *CoRR*, abs/1804.00610. Available at: http://arxiv.org/abs/1804.00610.

Nakamoto, S. (2008). Bitcoin: A peer-to-peer electronic cash system.

Nitti, M., Girau, R., Atzori, L., Iera, A. and Morabito, G. (2012). A subjective model for trustworthiness evaluation in the social internet of things. *2012 IEEE 23rd International Symposium on Personal Indoor and Mobile Radio Communications (PIMRC), IEEE*, Sydney, Australia, 18–23.

Schwartz, D., Youngs, N., Britto, A. *et al.* (2014). The ripple protocol consensus algorithm. Ripple Labs Inc, White Paper, 5.

Shah, S.H. and Yaqoob, I. (2016). A survey: Internet of Things (IOT) technologies, applications and challenges. *2016 IEEE Smart Energy Grid Engineering (SEGE)*, i, 381–385.

Sicari, S., Rizzardi, A., Grieco, L.A. and Coen-Porisini, A. (2015). Security, privacy and trust in internet of things: The road ahead. *Comput. Netw.*, 76, 146–164.

Stankovic, J.A. (2014). Research directions for the internet of things. *IEEE Internet of Things Journal*, 1(c), 3–9.

Wander, A.S., Gura, N., Eberle, H., Gupta, V., and Shantz, S.C. (2005). Energy analysis of public-key cryptography for wireless sensor networks. *IEEE Internet of Things Journal, 2005 (PerCom 2005), IEEE*, Kauai Island, HI, USA, 324–328.

Zamfir, V. (2017). Casper the friendly ghost, a correct-by-construction blockchain consensus protocol. Available at: https://github.com/ethereum/research/raw/master/papers/CasperTFG/CasperTFG.pdf.

Zhang, J. and Varadharajan, V. (2010). Wireless sensor network key management survey and taxonomy. *J. Netw. Comput. Appl.*, 33(2), 63–75.

Zhang, Z.K., Cheng Yi Cho, M., Wang, C.W., Hsu, C.W., Chen, C.K., and Shieh, S. (2014). Iot security: Ongoing challenges and research opportunities. *Proceedings – IEEE 7th International Conference on Service-Oriented Computing and Applications, SOCA*, 230–234.

7

How Machine Learning Can Help Resolve Mobility Constraints in D2D Communications

Chérifa BOUCETTA[1], **Hassine MOUNGLA**[1,2] **and Hossam AFIFI**[2]

[1]*LIPADE, University of Paris, France*

[2]*Telecom SudParis, Institut Polytechnique de Paris, Saclay, France*

7.1. Introduction

Device-to-Device (D2D) communications, as introduced by Release 12 of the 3rd Generation Partnership Project (3GPP), are based on geographic proximity. A user's equipment (UE) is able to detect and communicate with other UE in their neighborhood (Huo *et al.* 2017). D2D communications offer several advantages, such as optimal spectrum use and improvement in overall flow and latency. These communications are applied in different environments for both commercial, as well as public use.

The 3GPP standard defined several use cases in the field of public safety. Indeed, it is essential to maintain communications in environments that are characterized by no coverage or poor coverage, as in underground trains for instance, or in the case of large-scale natural disasters or even in the case of power blackouts. Thus, additional capacities must be provided when needed, especially in large metropolitan areas (Deng *et al.* 2015).

Service Level Management in Emerging Environments,
coordinated by Nader MBAREK.

Direct D2D discovery is one of the services that is jointly defined with D2D communication and D2D synchronization. This service allows a UE to identify nearby users and devices.

Two discovery models have been defined:

– model A is based on "announcements", where UE broadcasts information;

– model B is based on the request/response principle and is used when a UE requires specific information.

The base station for mobile networks based on Long-Term Evolution (LTE) technologies, called the evolved Node B (eNB), can play a role in direct D2D discovery, either by sharing the list of available resources or by managing the allocation of resources (Luo and Li 2012).

Thus, two types of resource allocation are defined: Type A and Type B. In Type B, called "scheduled" allocation, the D2D users are connected to the eNB that manages the resource allocation. In Type A, on the other hand, the parameters are preconfigured in the UE and this method is called "UE-selected". Both of these types of resource allocation are autonomous and independently select resources to exchange different messages, in order to carry out D2D discovery in discrete time intervals over a fixed duration (called the discovery period).

We can also define a third type of resource allocation based on the two types discussed above: UE recovers the pool of available resources from the eNB and each UE then selects the resources they require, independent of their neighbors (Deng *et al.* 2015).

In D2D discovery, and as with any other communication service, the variation in the number of users affects the performance of the network. Mobile users consume resources very differently based on their position, the type of applications and the network services offered. This results in large variations in usage profiles for the network and in user density profiles, based on each location.

In order to predict user density in real time and thus enhance the performance of the discovery process for dynamic environments, we propose a new approach in this chapter, which is based on machine learning tools: Support Vector Machines (SVM) and Support Vector Regression (SVR) (Basak *et al.* 2007). In the literature, the problem of searching for mobile networks and that of deep learning have generally been studied independently and it is only recently that the two fields have been examined together.

SVR is usually a machine learning approach that is used to process large datasets as it only uses a few samples (called support vectors) for all the learning data which significantly reduces processing time. Further, SVR is based only on the minimal calculation requirement, unlike other regression tools such as Artificial Neuronal Networks, which require greater processing time, more complex configuration and a lack of generality.

The proposed prediction model is based on real data provided by Orange Senegal. In an earlier study, we defined three categories of UE density profiles for the base station, based on the fluctuation model during the day. Each profile represents a specific use of resources. We then implemented a generator of realistic traffic based on density models that were already defined. This generator was finally used to predict the evolution of UE density in each zone and, consequently, to adapt the transmission probability in the process of discovering neighbors.

In this chapter, we will introduce D2D communications and research on future mobile networks in section 7.2. We will then examine machine learning techniques and the applications of ML in section 7.3. In section 7.4, we will describe the approach that we propose and in section 7.5, we will present the experimental results obtained. Finally, section 7.6 concludes the chapter and gives examples of future research to be carried out.

7.2. D2D communication and the evolution of networks

There are two distinct phases in D2D communication: a discovery phase, in which the D2D device detects compatible devices in its vicinity. There is then the second phase of data exchange, with the equipment configuring D2D links to send data.

In the literature (Doppler *et al.* 2009; Fodor *et al.* 2012; Lei *et al.* 2012) and as shown in Figure 7.1, two D2D modes have been proposed. In the distributed mode, the UE communicates directly with each other and are organized in a completely distributed manner. The network is therefore self-organized, without being controlled by the infrastructure.

In the second mode, called the Assisted Mode, network infrastructure is required to configure D2D communication. In sections 7.2.1 and 7.2.2, we will describe both phases of discovery and data communication that may be managed using one of the two models: either using a distributed approach or using a network assisted approach.

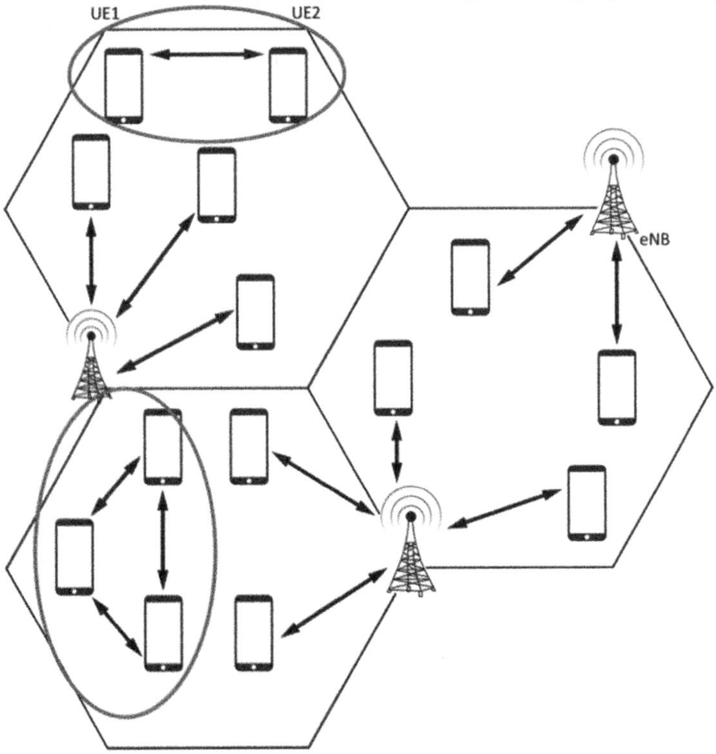

Figure 7.1. *Device-to-Device communication between UE1 and UE2 as a sublayer in a cellular network. For a color version of this figure, see www.iste.co.uk/mbarek/service.zip*

7.2.1. *The discovery phase in D2D communications*

The discovery phase may be centralized (based on network infrastructure) or distributed, also called direct discovery.

7.2.1.1. *The direct approach*

As per the 3GPP ProSe (3GPP 2014), two direct discovery models were identified: the first model (A) is the "I'm here" model based on the unconditional broadcast of a *Hello* message. The second model (B), however, is based on the request/response method. In effect, a UE device can use direct discovery in its environment, depending on the application and on the role of the discovery: for instance, in the discovery in active mode, the device announces its presence and the service it offers, while in the discovery of a surveillance application, the UE remains in passive mode and only monitors specific information in the fields of interest, that are announced in its vicinity.

The direct discovery approach presents several advantages, such as flexibility and scalability as it is based on the location of equipment and on local radio.

7.2.1.2. *The centralized approach*

This approach involves one or more network entities in the discovery process. As the network operators have a broader view of global traffic and the context of UE mobility, centralized discovery approaches aim to make use of the resources available to mobile operator networks on micro- and macromobility, in order to supply more precise information for detection. It must be noted that this discovery approach does not comply with the basic definition of D2D as communication without infrastructure between two devices, however this approach is still considered an alternative discovery model for proximity services, as per the 3GPP standard.

7.2.2. *The data exchange phase in D2D communications*

After the neighborhood discovery phase, the UE establishes communication links to exchange data. These links differ from traditional cellular network links and can be used for offloading. As shown in Figure 7.1, data can be transmitted either directly between D2D devices (direct transmission scheme) or via an optimized path, controlled by the operator through the eNB (controlled transmission system). Based on these two diagrams, communication between the devices can be carried out at as best as possible (no support for Quality of Service (QoS) and links without connection) or with QoS (setting up dedicated LTE data carriers) because the operator provides processing based on D2D services, data traffic flows, number of subscribers, etc.

In a conventional LTE communication (3GPP 2012; Jeong *et al.* 2013), the data flows between two devices are set up via the establishment of data carriers. A data carrier is composed of a radio support between the device and the eNB, an Evolved Packet Core (EPC) support between the eNB and the Core network, and packet filters on the devices. Setting up a data carrier consists of configuring a Packet Data Network (PDN) connection by specifying the PDN gateway and assigning an IP address to the communicating devices.

In the case of D2D communication, a data carrier need not be set up between the devices when the communication is through direct D2D offloading. Thus, data are exchanged directly over the radio link between the devices and only physical addressing is used to identify the source and the destination. In the case of a controlled offloading system, however, a dedicated D2D support, with specific radio resources allocated by the eNB, must be configured. In the latter case, the dedicated carrier comprises solely of a radio carrier without EPC support, where the data flows only pass between the eNB and the devices that use the same IP address as that assigned to their LTE supports, created during the first connection to the network.

7.2.3. *Investigations into future mobile networks*

The new generations of emerging mobile networks such as 5G and above integrate a host of new techniques in order to overcome the different performance limitations of current deployments, and to thus meet the new requirements of communication technologies. The progress made to date in this field has been summarized in several state of the art works, tutorials and magazine articles (Andrews *et al.* 2014; Gupta and Kumar Jha 2015; Agiwal *et al.* 2016; Panwar *et al.* 2016; Mao *et al.* 2017).

Andrews *et al.* highlight the differences between 5G and previous mobile network architectures, offer a comprehensive review of 5G techniques and discuss research problems that future developments must confront (Andrews *et al.* 2014). Agiwal *et al.* (2016) review new 5G network architectures, study emerging wireless technologies and flag unresolved research issues. Gupta and Kumar Jha (2015) also review existing work on 5G cellular network architectures, then propose a framework that integrates networking ingredients, such as D2D communication, small cells, cloud computing and the Internet of Things (IoT).

Smart mobile networking is becoming a popular research area and related work has been reviewed in the literature (e.g. Bkassiny *et al.* 2013; Abu Alsheikh *et al.* 2014; Buda *et al.* 2016; Keshavamurthy and Ashraf 2016; Bui *et al.* 2017; Jiang *et al.* 2017). Machine learning techniques could provide more solutions in this domain.

7.3. The context for machine learning and deep learning

Machine learning adds value to numerous fields of application, namely efficiency of systems/networks, such as energy and IT security (spam detection, threat analysis etc., Aono *et al.* 2018), image recognition (digital, satellite, handwriting), environmental management, IoT (Li *et al.* 2018), autonomous cars (Liu *et al.* 2017a) and voice recognition (Chen *et al.* 2014; Huo *et al.* 2017; Wu *et al.* 2017).

Machine learning techniques are essentially based on artificial intelligence tools (neural networks, decision trees and association rules) that aim to provide predictive analysis from a mass of heterogeneous data. In addition, these techniques allow correlations to be made between various events, in order to generate decisions. In other words, this involves defining a data model and making predictions using data based on collected statistics.

Recently, a category of machine learning methods that is based on deep artificial neural networks proved to be particularly well-suited to dealing with vast amounts of learning data and may prove to be significantly better than alternative methods. This category, called "Deep Learning", has been used in various areas of Artificial Intelligence (Hinton 2012; Xiaohui 2017).

Deep learning is based on a deep neural network (made up of tens or even hundreds of layers of neurons, each of which receives and interprets information from the previous layer) and a series of complex propositional calculations. Deep learning algorithms work with learning at multiple levels of detail. Thus, we move from low-level parameters to higher-level parameters through the different layers and levels corresponding to different levels of data abstraction. One extension of the neural network is Long Short Term Memory (LSTM), whose goal is to predict a future state based on previous states. The advantage of this model is the difference between the number of inputs and outputs.

There are different deep learning models in the literature, notably the Deep Belief Network (DBN) (Hinton *et al.* 2006; Le Roux and Bengio 2010), Stacked AutoEncoder (Vincent *et al.* 2010) and Convolutional Neural Networks (CNN) (Lecun *et al.* 1998). Recently, CNNs have proven to be a good alternative to other prediction-oriented deep learning models (Dieleman *et al.* 2015; Hosseini *et al.* 2017; Wang *et al.* 2017), as well as classification-oriented and detection-oriented models (Cao *et al.* 2018).

Machine learning uses the deep neural network to extract relevant characteristics from a mass of data, as shown in Figure 7.2. The deep neural network is the final stage of machine learning and the learning rule presents the algorithm that will generate the model from the learning data. These algorithms require increasing quantities of data to improve the efficiency of the learning process. In other words, the greater the volume of data (multimodal sources), the more efficient the process becomes. This process is therefore considered robust and high-performing, given its very high level of accuracy in optimizing and forecasting decisions.

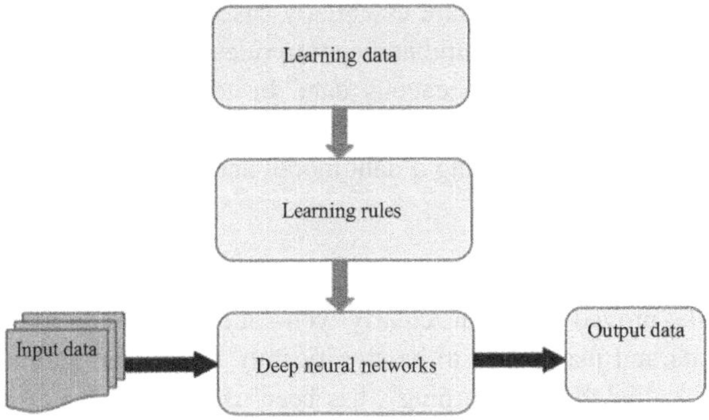

Figure 7.2. *The machine learning flowchart*

7.3.1. *Overview of deep learning and its application*

The age of Big Data has aroused a keen interest in deep learning in different research disciplines (Chen and Lin 2014; Najafabadi *et al.* 2015; Gheisari *et al.* 2017; Hordri *et al.* 2017). This has led to a rising number of publications on state-of-the-art and ongoing research (Deng 2014).

Lecun *et al.* (1998) offer an overview of deep learning, introducing several popular models and examining the potential of deep neural networks. Schmidhuber (2015), on the other hand, has undertaken an encyclopedic study of deep learning, which is likely to be the most comprehensive work on the subject to date. It covers the evolution, methods, applications and open research problems in this field.

Liu *et al.* (2017) summarize the principles underlying different deep learning models and examine evolutions in deep learning in various applications, such as speech processing, shape recognition and computer vision.

Zhang *et al.* (2017) study developments in deep learning for recommendation systems, which may play an important role in mobile advertising.

As deep learning becomes more and more popular, Goodfellow *et al.* (2016) offer a complete tutorial on it in a book that covers the prior knowledge required, underlying principles and well-known applications.

Another field of application for deep learning is communication within networks, which is where we position our approach, which uses machine learning in D2D discovery. In the following section, we introduce the different machine learning techniques and applications that allowed us to present our approach of dynamically adapting the transmission probability in D2D communication.

7.3.2. *Types of machine learning*

The main objective of machine learning is to find a model based on the collected data. It consists of extrapolating a law from all the experimental data while ensuring good quality. Datasets are divided into two categories in this context:

– *training set:* this is used to learn the model;

– *test set*: this is used to learn quality.

Among the typical problems addressed by machine learning are:

– regression: forecasting a value on the basis of other values;

– clustering: dividing things/people into natural groups;

– collaborative filtration: recommending articles to users based on past behaviors.

Machine learning techniques may be divided into three categories, based on the learning method:

– *supervised learning*: the training set contains the list of inputs with the correct output {input, correct output};

– *unsupervised learning*: the training set only contains inputs, without the correct output {input};

– *learning through reinforcement*: this is based on an input set, a few outputs and the quality of the result {input, correct output, grade for this output}.

Supervised learning is very similar to the manner in which humans learn things.

7.3.3. *Linear regression and classification*

We differentiate between two types of application of supervised learning: classification and regression.

Classification: this is based on the literal search for categories to which the learning data belongs and is presented in the following form: {input, class}.

Some examples of classification-based applications are:

– spam filter services → classifying emails into spam or not spam (regular emails);

– number recognition services → classification of images from 0 to 9;

– facial recognition services → classification of facial images.

Linear regression: the main goal of linear regression is to forecast values based on other values. For example, predicting a person's height based on their personal data (bank account, telephone number, address, etc.). Thus, the more data we have, the higher the chance of obtaining a model that is closer to reality.

The difference between classification and regression is that classification determines the group to which input data belongs. Following this, the correct output of the classification is given in categories. On the other hand, regression is based on prediction and considers values in the learning data to provide the correct output.

7.3.3.1. *Linear classification with SVM*

SVM belongs to the category of linear classifiers (which use linear separation of data) and has its own method for finding the boundary between categories. This technique uses geometric criteria. Let us assume that we have a set of training samples containing vectors in a d-dimensional space: ($x_i \in R^d$, $i = 1,2,\ldots,N$), with a class label $y_i \in \{-1,1\}$. Thus, the SVM classifier (Hearst *et al.* 1998) is represented by a function $f(x, \alpha) \rightarrow y$, where α is the parameter of the classifier.

In general, the SVM approach consists of searching for a hyperplane that maximizes the distance between the training samples and the separation hyperplane, as given below:

– samples labeled $y_i \in \{-1,1\}$ are localized to each side of the hyperplane;

– the hyperplane must maximize the *margin*, i.e. the minimum distance between the training vectors and the hyperplane. These vectors are called the *support vectors*.

The hyperplane is defined by $w \cdot x + b = 0$, where w and b are the hyperplane parameters. The vectors that are not in this hyperplane are determined using: $w.x + b \geq 0$ and hence the SVM is defined as follows:

$$f(x, \alpha) = \operatorname{sgn}(w.x + b)$$

In order to find a hyperplane, the parameters b and w must be estimated using equation [7.1]:

$$y_i(w.x_i + b) > 0, \text{where } i = 1,2, \ldots, N \qquad [7.1]$$

The distance between the closest training sample and the hyperplane separator may be equal to $\frac{1}{\|w\|}$, with a simple scaling of the hyperplane parameters w and b such that:

$$\min\left(y_i(w.x_i + b)\right) \geq 1, \text{where } i = 1,2, \ldots, N \qquad [7.2]$$

The geometric margin between the two classes is given by the quantity. Consequently, the maximization of the margin leads to an optimization problem with the following constraint:

$$\begin{Bmatrix} \text{minimizing: } \frac{1}{\|W\|^2} \\ \text{the subject of: } y_i(w.x_i + b)) \geq 1, \text{where } i = 1,2, \ldots, N \end{Bmatrix} \qquad [7.3]$$

7.3.3.2. *Support Vector Regression*

SVR uses the same principles as SVM for classification, but with a few differences. As can be seen from its name, SVR uses continuous values instead of classification in SVM. The principle here is to find the "flattest" possible linear regression line, subject to certain constraints. However, the main idea is to minimize error and individualize the hyperplane that maximizes the margin. Thus, a tolerance margin (epsilon) is defined as being an approximation of the SVM.

7.4. Dynamic discovery

In this section, we will describe a new approach based on machine learning tools, in particular, SVM and SVR, which can enhance the performances of the discovery process for dynamic environments. We will consider the direct discovery approach, given the advantages that this offers compared to the centralized approach. However, the principal challenge for D2D discovery is the increase in the number of sent messages during the discovery and surveillance procedures. The discovery of UE, in particular, can consume a lot of energy due to the continuous, periodic transmission of discovery messages. These machine learning algorithms mainly aim to predict the density of users in the cells, in order to reduce the number of messages sent and the energy consumed. Indeed, the solution allows the UE to modify their transmission probabilities as needed, in order to reduce the time required to discover other UE. Thus, this makes it possible to enhance and maintain network performances or even improve them during the discovery phase, in scenarios outside network coverage

7.4.1. *Real-time prediction of user density*

We will define an SVR-based model to predict user density within cells (Hammami 2018). Given a sample of N time series (the history of UE density for the concerned cells) denoted by $e(x_i, y_i)^N$ where SVR aims to find a linear function that associates x_i and y_i in a function space (of higher dimension). In our proposition, we use the Radial Basis Function (RBF) as the kernel function as it is relevant to nonlinear time series (as in our case) due to its capacity for generalization and its capacity for nonlinear mapping in an infinite function space.

RBF networks are connectionist models that are easy to implement and are quite intelligible. They are very widely used for regression and discrimination. Their theoretical and practical properties have been studied in great detail in the literature. Table 7.1 describes a few examples of kernel functions used in the literature.

Kernel functions	Equation
Linear	$K(x_i, x_j) = < x_i . x_j > +b$
Polynomial	$K(x_i, x_j) = < x_i . x_j >^d + b$
Sigmoid	$K(x_i, x_j) = \text{tang}(< x_i . x_j >) + b$
RBF	$K(x_i, x_j) = \exp(-\gamma \|x_i - x_j\|)^2$

Table 7.1. *Examples of SVR kernel functions*

The prediction is made in two stages: the first step is training, where we analyze and use historical data on user density; the second step is testing, where the model's parameters are optimized.

The efficiency of SVR prediction is validated using test data taken from the Orange Dakar dataset. The effectiveness of the SVR-based prediction is measured using two major indicators: the squared correlation coefficient, as well as the Mean Squared Error (MSE). Prior to this research, in the context of another study (Hammami *et al.* 2016) based on a large collection of data from Orange Senegal, we deduced and categorized three UE density profiles for the selected base stations extracted. Each type of base station profile is associated with a specific load fluctuation and network use model that depends on the nature of its position. The classification of the profile for the base station in Dakar has also been taken into account. The "night load" profile represents residential zones, while the "day load" profile represents most sectors of activity and the "constant load" profile is associated with mixed zones. A collection of UE density models is deduced from the Dakar data to be incorporated in ns-3 in order to implement a realistic traffic generator. The tool is used to predict the evolution of UE density in each zone and to then use the predicted values for the direct discovery algorithm.

7.4.2. *The dynamic discovery algorithm*

Direct neighborhood discovery in D2D communications suffers from a lack of flexibility, with respect to the ambient communication environments, because it is based on predefined parameters, especially for transmission probability. The 3GPP standard stipulates that the channel used for discovery must be mi-duplex and the set of UE that use the same transmission probability that is pre-configured and defined in the pool of resources for discovery in the mode selected by the UE. However, the use of adaptive transmission probability values, selected in real-time based on the size of the group, could lead to significant improvement in the performance of the process overall.

The optimal transmission probability, θ, is thus calculated as (Ben Mosbah 2016):

$$\theta = \begin{cases} 1 & \text{If } N_u < \frac{N_r(N_t-2)+N_t}{N_t-1} \\ \frac{2N_r+N_t(N_u-1)-\sqrt{4N_r(N_r-N_t)+N_t^2(N_u-1)^2}}{2\,N_u} & \text{else} \end{cases} \qquad [7.4]$$

We propose an algorithm that uses the predicted user density models in order to determine the value of the transmission probability θ, that offers the best discovery performance.

The defined algorithm is run on an hourly basis and takes into account the predicted number of users N_u for this zone, for the corresponding period of the day. Using [7.4], we can calculate the optimal transmission probability value in real-time discovery. For each period (the start of every hour, for instance), the algorithm considers the geographic zone to determine the type of traffic. The algorithm consults its database, made up of actual logging of similar traffic, and initiates computation of the average, obtaining the predicted number of users for this zone, for this time and for this specific type of traffic. It must be noted that the user density profiles used during this process are extracted from real data in the network. Consequently, this gives us an accurate idea of the real number of UE present in the network. Based on this information, the optimal transmission probability θ is periodically computed on an hourly basis (the predicted UE number may change for each period) in order to obtain the best discovery performance. We have demonstrated the effectiveness of this proposition through significant reductions in energy consumption and discovery time.

7.5. Experimental results

In this section, we discuss the evaluations of the proposed learning method using simulations under ns-3 (Rouil *et al.* 2017). We use a scenario taken from the study of the traces of the operator Orange Senegal (Hammami *et al.* 2016) that we integrated into ns-3 as realistic UE density models, in order to study the effectiveness of our proposed approach.

7.5.1. *General hypotheses*

Table 7.2 summarizes the list of network parameters as well as the corresponding values used in our simulations. The sets of UE models are taken from the Dakar data so that they can be integrated into ns-3, in order to implement a tool to generate realistic traffic. The tool is then used to generate a realistic model in order to train the SVR model and predict the evolution of the UE density in each zone (Ben Mosbah 2016). This is done so that the predicted values can be used for the direct D2D discovery algorithm.

Parameter	Value
Transmission power of the UE	23 dBm
Propagation model used	Cost231 (Damasso 1999)
Available bandwidth	50 RB
Carrier frequency	700 MHz
Discovery period	0.32 s
Number of retransmissions	0
Number of repetitions	1
Number of pairs of resource blocks	4
Number of subframes	5
Total number of resources	20
Total number of eNBs	3
Start of discovery	2
Total number of simulations per scenario	100

Table 7.2. *Simulation parameters and values*

We considered two scenarios with low density and high density for the "day charge" profile. These are presented in sections 7.5.2 and 7.5.3.

7.5.2. *Traffic with low user density*

We consider a scenario for a user density profile in the morning in a zone characterized by fluctuations in use, due to the presence of higher education institutions. Consequently, a profile for morning traffic is taken into account. Under normal conditions, the zone would be served by three eNBs, however, in this scenario, the eNBs have been made unusable (due to a loss in backhaul connectivity, for example). The user equipments are randomly deployed, as illustrated in Figure 7.3. The maximum number of users at peak hours is 50, 75 and 100.

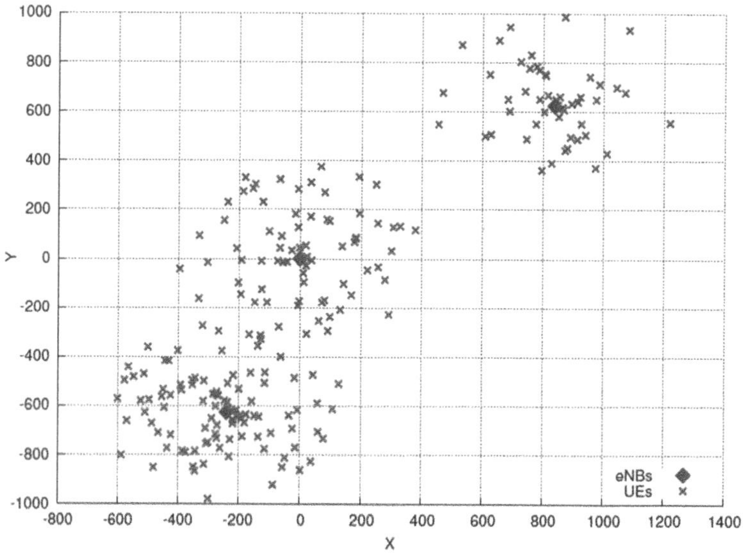

Figure 7.3. *Example of node positions for a morning traffic scenario with low user density. For a color version of this figure, see www.iste.co.uk/mbarek/service.zip*

7.5.3. *Traffic with high user density*

Contrary to the scenario described in the previous section, consider a dense geographic zone (more precisely, a business district in Dakar) covered by 13 eNBs, as shown in Figure 7.4. This zone could have a traffic profile with a higher concentration of deployed UE.

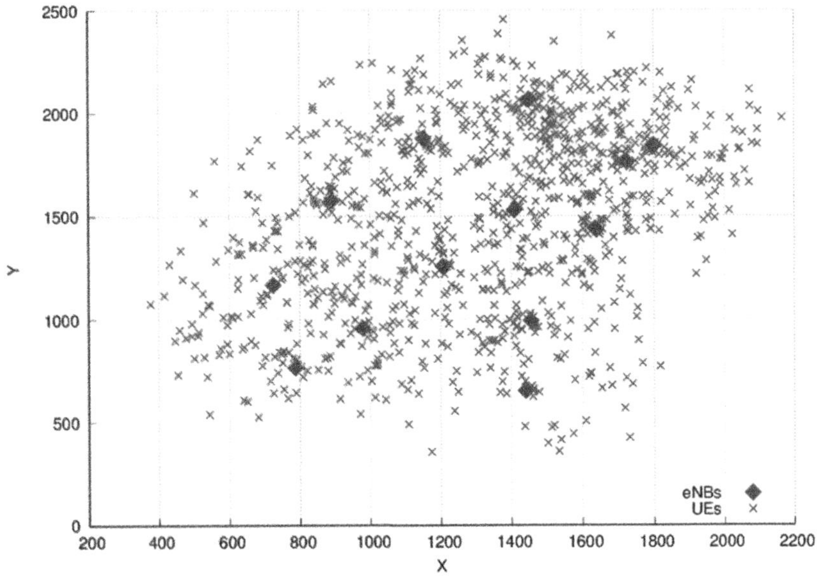

Figure 7.4. *Example of the node positions for a morning traffic scenario with high user density. For a color version of this figure, see www.iste.co.uk/mbarek/service.zip*

The simulation results are illustrated in Figures 7.4–7.6. Each of these figures indicates the number of messages and duration (measured in period) until all UE in the zone are discovered (or until the 60 second discovery period expires). In the 3GPP algorithm, the transmission probability is invariable for the entire simulation and may be equal to 0.25, 0.5, 0.75 or 1. Despite this, however, our algorithm uses information related to the predicted number of users, to modify the corresponding transmission probability in a dynamic manner. The UE can calculate their own transmission probability over time. The planned discovery method makes it possible to reduce energy consumption and prevent large packet loss because of the reduction in the number of messages transmitted.

We can observe an increase in network activity, which reaches its peak toward noon. The curves testify to the fact that we are simulating morning traffic. It can also be seen that during the hours when the number of UE is low, there is a direct compromise between the time required for discovery and the number of messages sent. We can see how our algorithm establishes an equilibrium between these two measures.

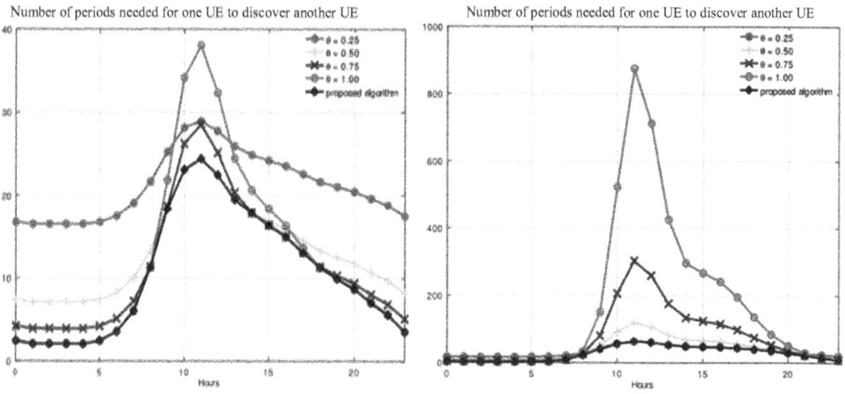

(a) *Low density case* (b) *High density case*

Figure 7.5. *Number of periods required for complete discovery. For a color version of this figure, see www.iste.co.uk/mbarek/service.zip*

With respect to the messages, only the configuration with a transmission probability of 0.25 offered the best performances at certain times and at these times the time metric was the best for our algorithm.

A factor that may prove more important is that our algorithm showed more predictable behavior (with more coherent values for a certain number of messages and time required) than the 3GPP algorithms.

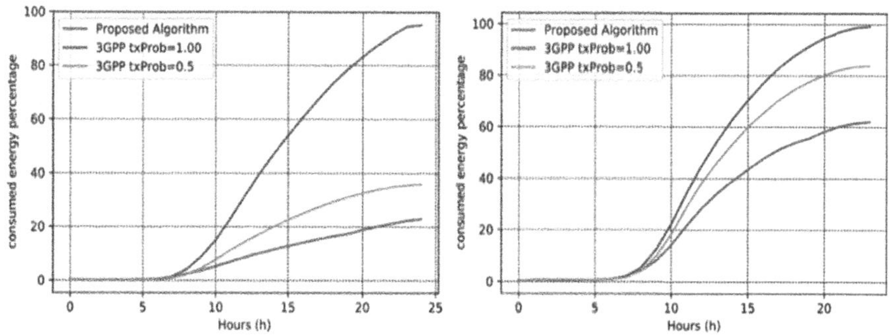

(a) *Low density case* (b) *High density case*

Figure 7.6. *Percentage of energy consumed for the two cases considered: low and high density. For a color version of this figure, see www.iste.co.uk/mbarek/service.zip*

7.6. Conclusion

In this chapter, we proposed a model based on machine learning in the context of D2D communication, with the goal of optimizing the use of network resources.

We took into account the history of user density using real operator data, with the aim of optimizing the discovery process by defining an optimized transmission probability for each discovery configuration. We used data from real networks and proposed a proactive algorithm that predicted the number of users in a specific zone and proactively adjusted the transmission probability, in order to optimize transmissions for higher efficiency.

User density models were first extracted and synthesized, varying from one zone to another, depending on their geographic location, as well as their type of traffic (i.e. industrial, residential and mixed). These models were used to create a history of the number of users located around a zone for every hour in the day. These density models were then integrated into our algorithm in order to calculate the optimized transmission probability for each period.

It was seen that machine learning could achieve these objectives with greater efficiency than conventional models/algorithms, whose accuracy could be limited.

Further, D2D communications are still being standardized. Vehicular communications (V2X) are a part of this and were recently introduced (3GPP, Release 14) in the context of D2D communications, in order to allow vehicles to communicate effectively (vehicle-to-vehicle, for instance, or vehicle-to-infrastructure and pedestrians). This may result in stricter requirements in terms of delays and transmission loss due to the high mobility of vehicular networks.

More advanced algorithms can also be designed using advanced machine learning, coupled with a large-scale analysis of data, such that the algorithm itself evolves over time to reflect the various behavioral changes in its users. In this regard, we wish to draw particular attention to multilayer deep learning.

7.7. References

3GPP (2012). Feasibility study for evolved Universal Terrestrial Radio Access (UTRA) and Universal Terrestrial Radio Access Network (UTRAN). Report, 3GPP, USA.

3GPP (2014). Architecture enhancements to Support Proximity Services (ProSe). Report, 3GPP, USA.

Abu Alsheikh, M., Lin, S., Niyato, D., and Tan, H.P. (2014). Machine learning in wireless sensor networks: Algorithms, strategies, and applications. *IEEE Communications Surveys & Tutorials*, 16(4), 1996–2018.

Agiwal, M., Roy, A., and Navrati Saxena, N. (2016). Next generation 5G wireless networks: A comprehensive survey. *IEEE Communications Surveys & Tutorials*, 18(3), 1617–1655.

Andrews, G.F., Buzzi, S., Choi, W., Hanly, S.V., Lozano, A., Soong, A., and Zhang, J.C. (2014). What will 5G be? *IEEE Journal on Selected Areas in Communications*, 32(6), 1065–1082.

Aono, Y., Hayashi, T., Wang, L., Phong, L.T., and Moriai, S. (2018). Privacy-preserving deep learning via additively homomorphic encryption. *IEEE Transactions on Information Forensics and Security*, 13(5), 1333–1345.

Basak, D., Pal, S., and Patranbis, D.C. (2007). Support vector regression. *Neural Information Processing-Letters and Reviews*, 11(10), 203–224.

Ben Mosbah, A. (2016). Context-aware mechanisms for device discovery optimization. PhD Thesis, Telecom SudParis and Pierre and Marie Curie University, Paris.

Bkassiny, M., Li, Y., and Jayaweera, S.K. (2013). A survey on machine-learning techniques in cognitive radios. *IEEE Communications Surveys & Tutorials*, 15(3), 1136–1159.

Buda, T.S., Assem, H., Xu, L., Raz, D., Margolin, U., Rosensweig, E., Lopez, D.R., Corici, M.I., Smirnov, M., and Mullins, R. (2016). Can machine learning aid in delivering new use cases and scenarios in 5G? *Network Operations and Management Symposium (NOMS)*. IEEE/IFIP, 1279–1284.

Bui, N., Cesana, M., Hosseini, S.A., Liao, Q., Malanchini, I., and Widmer, J. (2017). A survey of anticipatory mobile networking: Context-based classification, prediction methodologies, and optimization techniques. *IEEE Communications Surveys & Tutorials*, 19(3), 1790–1821.

Cao, X., Zhou, F., Xu, L., Meng, D., Xu, Z., and Paisley, J. (2018). Hyperspectral image classification with Markov random fields and a convolutional neural network. *IEEE Transactions on Image Processing*, 27(5), 2354–2367.

Chen, X. and Lin, X. (2014). Big data deep learning: Challenges and perspectives. *IEEE Access*, 2, 514–525.

Chen, Y., Lin, Z., Zhao, X., Wang, G., and Gu, Y. (2014). Deep learning-based classification of hyperspectral data. *IEEE Journal of Selected Topics in Applied Earth Observations and Remote Sensing*, 7(6), 2094–2107.

Damasso, E. (1999). Digital mobile radio towards future generation systems: COST Action 231. Report, European Commission, Brussels.

Deng, L. (2014). A tutorial survey of architectures, algorithms, and applications for deep learning. *APSIPA Transactions on Signal and Information Processing*, 3(2), 1–29.

Deng, L., Zhang, Y., Chen, M., Li, Z., Lee, J.Y., Zhang, Y.J., and Song, L. (2015). Device-to-device load balancing for cellular networks. *IEEE 12th International Conference on Mobile Ad Hoc and Sensor Systems (MASS)*. IEEE, 19–27.

Dieleman, S., Willet, K.W., and Dambre, J. (2015). Rotation-invariant convolutional neural networks for galaxy morphology prediction. *Monthly Notices of the Royal Astronomical Society*, 450(2), 1441–1459.

Doppler, K., Rinne, M., Wijting, C., Ribeiro, C.B., and Hugl, K. (2009). Device-to-device communication as an underlay to LTE-advanced networks. *IEEE Communications Magazine*, 47(12), 42–49.

Fodor, G., Dahlman, E., Mildh, G., Parkvall, S., Reider, N., Miklos, G., and Turanyi, Z. (2012). Design aspects of network assisted device-to-device communications. *IEEE Communications Magazine*, 50, 170–177.

Gheisari, M., Wang, G., and Bhuiyan, M.Z.A. (2017). A survey on deep learning in big data. *IEEE International Conference on Computational Science and Engineering (CSE) and Embedded and Ubiquitous Computing (EUC)*. IEEE, 2, 173–180.

Goodfellow, I., Bengio, Y., and Courville, A. (2016). *Deep Learning*. MIT Press, Cambridge, Massachusetts.

Gupta, A. and Kumar Jha, R. (2015). A survey of 5G network: Architecture and emerging technologies. *IEEE Access*, 3, 1206–1232.

Hammami, S.E. (2018). Dynamic network resources optimization based on machine learning and cellular data mining. PhD Thesis, Telecom Sud, Paris.

Hammami, S.E., Afifi, H., Marot, M., and Gauthier, V. (2016). Network planning tool based on network classification and load prediction. *Wireless Communications and Networking Conference (WCNC)*. IEEE, 1–6.

Hearst, M.A., Dumais, S.T., Osuna, E., Platt J., and Scholkopf, B. (1998). Support vector machines. *IEEE Intelligent Systems and Their Applications*, 13(4), 18–28.

Hinton, G.E., Krizhevsky, A., and Sutskever, I. (2012). Imagenet classification with deep convolutional neural networks. *NIPS'12 Proceedings of the 25th International Conference on Neural Information Processing Systems, Volume 1*, 1097–1105.

Hinton, G.E., Osindero, S., and Teh, Y.-W. (2006). A fast learning algorithm for deep belief nets. *Neural Computation*, 18(7), 1527–1554.

Hordri, N., Samar, N., Yuhaniz, S., and Shamsuddin, S.M. (2017). A systematic literature review on features of deep learning in big data analytics. *International Journal of Advances in Soft Computing & Its Applications*, 9(1), 32–49.

Hosseini, M.-P., Pompilli, D., Elisevich, K., and Soltanian-Zadeh, H. (2017). Optimized deep learning for EEG big data and seizure prediction BCI via Internet of Things. *IEEE Transactions on Big Data*, 3(4), 392–404.

Huo, Y., Dong, X., and Xu, W. (2017). 5G cellular user equipment: From theory to practical hardware design. *IEEE Access*, 5, 13992–14010.

Jeong, Y., Soon, Y.L., Jun Park, H., and Park, N.H. (2013). Solving the data overload: Device-to-device bearer control architecture for cellular data offloading. *IEEE Vehicular Technology Magazine*, 8(1), 31–39.

Jiang, C., Zhang, H., Ren, Y., Han, Z., Chen, K.C., and Hanzo, L. (2017). Machine learning paradigms for next generation wireless networks. *IEEE Wireless Communications*, 24(2), 98–105.

Keshavamurthy, B. and Ashraf, M. (2016). Conceptual design of proactive SONs based on the big data framework for 5G cellular networks: A novel machine learning perspective facilitating a shift in the SON paradigm. *System Modeling & Advancement in Research Trends (SMART), International Conference.* IEEE, 298–304.

Le Roux, N. and Bengio, Y. (2010). Deep belief networks are compact universal approximators. *Neural Computation*, 22(8), 2192–2207.

Lecun, Y., Bottou, L., Bengio, Y., and Haffner, P. (1998). Gradient-based learning applied to document recognition. *Proceedings of the IEEE*, 86(11), 2278–2324.

Lei, L., Zhong, Z., Lin, C., and Shen, X. (2012). Operator controlled device-to-device communications in LTE-advanced networks. *IEEE Wireless Communications*, 19, 96–104.

Li, H., Ota, K., and Dong, M. (2018). Learning IoT in edge: Deep learning for the Internet of Things with edge computing. *IEEE Network*, 32(1), 96–101.

Liu, S., Tang, J., Wang, C., Wang, Q., and Gaudiot, J.-L. (2017a). A unified cloud platform for autonomous driving. *Computer*, 50(12), 42–49.

Liu, W., Wang, Z., Liu, X., Zeng, N., Liu, Y., and Ealsaadi, F. (2017b). A survey of deep neural network architectures and their applications. *Neurocomputing*, 234, 11–26.

Luo, J. and Li, X. (2012). Physical downlink control channel format selection of evolved node B in long term evolution system. *5th International Conference on Biomedical Engineering and Informatics (BMEI)*. IEEE, 1494–1498.

Mao, Y., You, C., Zhang, J., Huang, K., and Letaief, K.B. (2017). A survey on mobile edge computing: The communication perspective. *IEEE Communications Surveys & Tutorials*, 19(4), 2322–2358.

Najafabadi, M., Villanustre, F., Khoshgoftaar, T., Seliya, N., Wald, R., and Muharemagic, E. (2015). Deep learning applications and challenges in big data analytics. *Journal of Big Data*, 2(1), 1.

Panwar, N., Sharma, S., and Kumar Singh, A. (2016). A survey on 5G: The next generation of mobile communication. *Physical Communication*, 18, 64–84.

Rouil, R., Cintron, F.J., Benmosbah, A., and Gamboa, S. (2017). Implementation and Validation of an LTE D2D Model for ns-3. *Proceedings of the Workshop on ns-3*. ACM, 55–62.

Schmidhuber, J. (2015). Deep learning in neural networks: An overview. *Neural Networks*, 61, 85–117.

Vincent, P., Larochelle, H., Lajoie, I., Bengio, Y., and Manzagol, P.-A. (2010). Stacked denoising autoencoders: Learning useful representations in a deep network with a local denoising criterion. *Journal of Machine Learning Research*, 11, 3371–3408.

Xiaohui, L., Liu, W., Wang, Z., Zeng, N., Liu, Y., and Alsaadi, F.E. (2017). A survey of deep neural network architectures and their applications. *Neurocomputing*, 234, 11–26.

Wang, L., Zhang, L., and Yi, Z. (2017). Trajectory predictor by using recurrent neural networks in visual tracking. *IEEE Transactions on Cybernetics*, 47(10), 3172–3183.

Wu, B., Li, K., Ge, F., Huang, Z., Yang, M., Siniscalchi, S.M., and Lee, C.H. (2017). An end-to-end deep learning approach to simultaneous speech dereverberation and acoustic modeling for robust speech recognition. *IEEE Journal of Selected Topics in Signal Processing*, 11(8), 1289–1300.

Zhang, S., Yao, L., Sun, A., and Tay, Y. (2019). Deep learning based recommender system: A survey and new perspectives. *ACM Computing Surveys*, 52(1), 5:1–5:38.

8

The Impact of Cognitive Radio on Green Networking: The Learning-through-reinforcement Approach

Mohammed Salih BENDELLA[1] **and Badr BENMAMMAR**[2]

[1]*STIC, University of Abou Bekr Belkaïd, Tlemcen, Algeria*

[2]*LTT, University of Abou Bekr Belkaïd, Tlemcen, Algeria*

8.1. Introduction

While current Information and Communication Technology (ICT) is extremely advanced, using this technology has a high fossil fuel cost and we must consider the environmental damage resulting from the CO_2 emissions of these technologies. There are three major kinds of pollution resulting from the field of communication technology: dissipation of energy, electromagnetic radiation and waste. The energy impact of ICT is estimated to account for about 2% of all greenhouse gas emissions (equivalent to total global emissions due to aviation) (TC Group 2008).

Various industries have expressed an interest in cutting down their energy consumption for economic reasons as well as for marketing purposes. Multiple sectors that work in a variety of fields have rushed forward to take up this cause, including electronics designers, the Information and Communications technology sector and, above all, the networking sector. Data centers and network infrastructure,

Service Level Management in Emerging Environments,
coordinated by Nader MBAREK. © ISTE Ltd 2020.

for example, require high performance and a high availability of machines. They draw energy from powerful devices that require constant air conditioning to operate (leading to a very high consumption of energy) and that are organized in a redundant architecture.

In recent times, a lot of productive effort and research has gone into cutting down on energy waste, a process that was described as making technology and network protocols more eco-friendly. These efforts led to the emergence of the concept of green networking, a movement that will transform ICT processes.

Green networking is a modern and fledgling concept that refers to the processes used to optimize a network in order to make it more energy efficient. It includes devices that make it possible to reduce energy consumption, which will ultimately minimize the volume of CO_2 emissions from telecommunication networks.

8.2. Green networking

8.2.1. *Why should we reduce energy consumption?*

Our planet is undergoing dangerous changes to which humans are only gradually paying heed. These are environmental problems related to the greenhouse gas effect. The presence of these gases in the atmosphere has increased significantly over the past few decades. The disastrous consequences of this phenomenon are being increasingly discussed today, with many studies focusing on the devastating effects of massive emissions of greenhouse gases and how they will impact climate change. In Europe, according to Bernstein *et al.* (2008) emissions need to decrease by 80–95% by 2050 to restrict the rise in global temperatures to under 2°C. However, as research in numerous studies has shown, ICT remains a significant source of energy consumption and greenhouse gas emissions. An innovative approach here is the possibility of making network devices and protocols more aware of the energy they consume, in order to make them energy efficient.

8.2.2. *Where can we reduce energy consumption?*

Before trying to reduce energy consumption, or even trying to understand how this can be done, we must identify where the most effective improvements can be made.

In 2002, Roth *et al.* (2002) studied the energy consumption of various categories of equipment in the global Internet network. The numbers from their study, given in Figure 8.1, show that local networks, due to their hubs and switches, are responsible for around 80% of the Internet's total energy consumption.

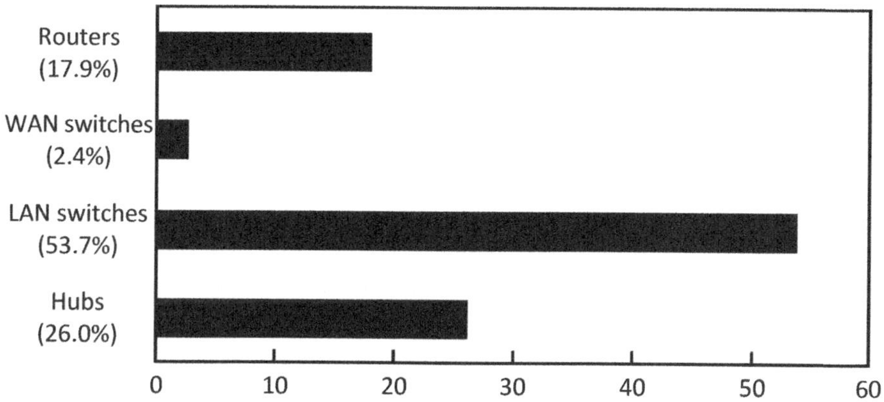

Figure 8.1. *The contribution of different devices to the network's energy consumption (Roth et al. 2002)*

In 2005, Nordman and Christensen (2005) studied the contribution of Network Interface Cards (NIC) to energy consumption, as well as the contribution of all other network elements, and then deduced that NICs are responsible for close to half the total energy consumption of a network.

Of course, ICT fields see very rapid changes, which could lead to inaccurate estimates. Energy consumption must, therefore, be constantly evaluated in order to update the most relevant targets for any potential reduction in energy waste.

8.2.3. *Definition and objectives of green networking*

Green networking uses processes that optimize a network in order to make it more energy efficient. Thus, it includes procedures that lead to a reduction in energy consumption, which, in turn, will lead to a reduction in the volume of CO_2 and other greenhouse gases emitted by telecommunications infrastructure. From a technical point of view, it is essential to maintain the same level of performance while also carrying out a given task, and we must put in place a green network that can do this. From a purely environmental perspective, the main objective of a green network is to minimize the emission of greenhouse gases. In order to achieve this goal, renewable energy must be used as widely as possible in ICT. Another eco-friendly step is to develop low-power components that are able to provide the same level of performance.

8.3. Green strategies

Traditionally, the form and size of network systems and network themselves have been based on principles that are inherently in opposition to the objectives of a green network, namely oversupplying, redundancy and tolerance to failure. Network devices are added to the infrastructure in order to take over when another machine malfunctions, which adds to the network's overall energy consumption.

Bianzino *et al.* (2012) demonstrate the key paradigms shifts that network infrastructure can put in place in order to achieve the green objectives listed above. They define four categories of solutions: *consolidation of resources*, *virtualization*, *selective connectivity*, and *energy-proportional computing*. These four categories represent four avenues of research that can be explored to improve device design and protocols.

8.3.1. *Consolidation of resources*

This includes all dimensioning strategies that can minimize worldwide consumption due to equipment that is being underutilized at any given time. Since the volume of traffic on a given network follows an approximate daily pattern and weekly behavior that are known (Qureshi *et al.* 2009), there is a clear opportunity to adapt the level of active oversupplying to the actual present requirements of a network. In other words, the required performance can always be guaranteed, but using a quantity of resources that is adapted to the actual demands of network traffic at that point in time, rather than resources being matched to peak hour traffic at all times. This can be done by switching off routers with a light load and by diverting traffic to a smaller number of pieces of active network equipment. The consolidation of resources is already quite a popular approach in other domains, especially in data centers.

8.3.2. *Selective connectivity*

Applying the same concept, according to Christensen *et al.* (2004) and Allman *et al.* (2007), selective connectivity consists of distributed mechanisms that allow a single piece of equipment to go into "sleep mode" for several hours, in a manner that is as transparent as possible to the other network administrative processes. While the consolidation of resources is applied to resources that are shared within the network infrastructure, selective connectivity makes it possible to block resources that are not being used on the network. For example, peripheral nodes may remain inactive in order to avoid supporting network connectivity tasks (such as sending periodic pulse signals, receiving radio-broadcast traffic, etc.). These tasks may be taken on by other nodes who temporarily also take on the identity of the inactive nodes, such that no fundamental change needs to be made to network protocols.

8.3.3. *Virtualization*

Virtualization covers a group of mechanisms that makes it possible for more than one service to be operated on the same piece of hardware, thereby improving its use. This results in lower energy consumption as a single machine with a heavy load consumes less energy than multiple machines with lighter loads, which is quite often the case. Virtualization may be applied to multiple resources, including network links, storage procedures, software resources, etc. A typical example of virtualization is sharing servers in data processing centers, which brings down hardware costs, improves energy management and minimizes the energy and financial costs of cooling – all of which result in a smaller carbon footprint for the data processing center.

Virtualization has already been successfully deployed: for example, the United States Postal Service virtualized 791 of their 895 physical servers (EPA 2007). Given that virtualization is a more mature sphere of research, any reader who wishes to learn more about this can refer to Nanda and Chiueh (2005) for a detailed but concise introduction to virtualization techniques in terms of IT architecture, while Chowdhury and Boutaba (2008) offer an introduction in terms of network management. At the same time, we still do not have a virtualization solution designed explicitly for reducing a network's energy consumption.

8.3.4. *Energy-proportional computing*

This was embedded (Barroso and Holzle 2007) through adaptation – it can be applied to an entire system, to network protocols as well as to devices and individual components. This principle can be illustrated as shown in Figure 8.2, representing the diverse energy-consumption (or cost) profiles of a device, depending on how much it is being used (or depending on the demand).

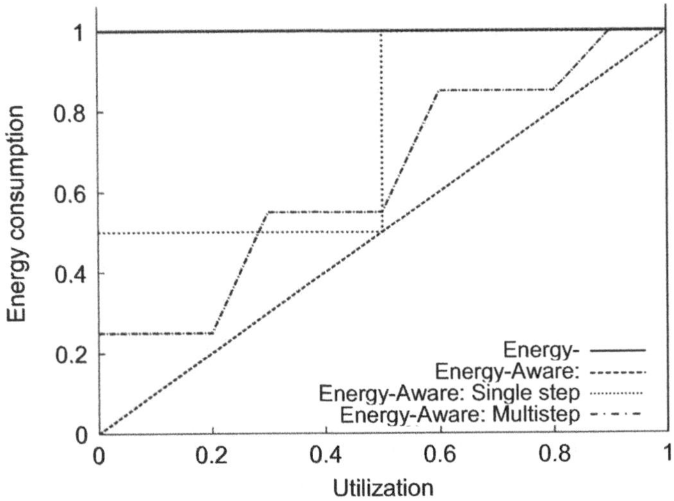

Figure 8.2. *Energy consumption based on use (Barroso and Holzle 2007)*

The metrics for utilization and energy have been normalized and the axis goes from 0 to 1. These different profiles offer various opportunities for optimization. Energy-agnostic devices, for which energy consumption is constant regardless of how much they are used, represent the worst cases: these devices are either powered on and consume the maximum energy or are switched off and non-operational.

However, devices that are very energy-aware are able to consume energy proportional to their utilization levels. While energy-agnostic devices and energy-aware devices represent two extremes of the possible profiles, there are an infinite number of intermediate profiles possible, for example the single step and multistep cases, represented in Figure 8.2, for which energy consumption adapts itself, approximately, to the load on the device. Single step devices have two operating modes, while multistep devices have different thresholds of performance.

Dynamic Voltage Scaling and Adaptive Link Rate are commonly used models in energy-proportional computing. Dynamic Voltage Scaling (Weiser *et al.* 1994) minimizes the CPU's energy state based on the load on the system. Adaptive Link Rate takes into account the load on a link and applies a similar concept to the network interfaces, reducing their capacity and, therefore, their consumption.

An additional problem that must still be resolved, however, is the selection and application of target loads in a distributed architecture, as the overall consumption of energy depends on the utilization level at which any energy-aware device is operating. For example, Figure 8.3 represents an energy-proportional carbon footprint as well as two others. One of these is better-suited to devices/protocols normally functioning at higher utilization levels, while the other is more effective for lower operational loads. Systems may be motivated to adopt this latter model due to the fact that data centers and network backbones are estimated to be functioning at low utilization levels most of the time. On the other hand, according to the principle of consolidation of resources, future systems will be incentivized to function at higher loads. This kind of interaction brings in another interesting dimension to consider: we will not only be designing adaptive systems based on various levels of utilization, but it will be much more apt to design systems optimized for the most *probable* utilization level.

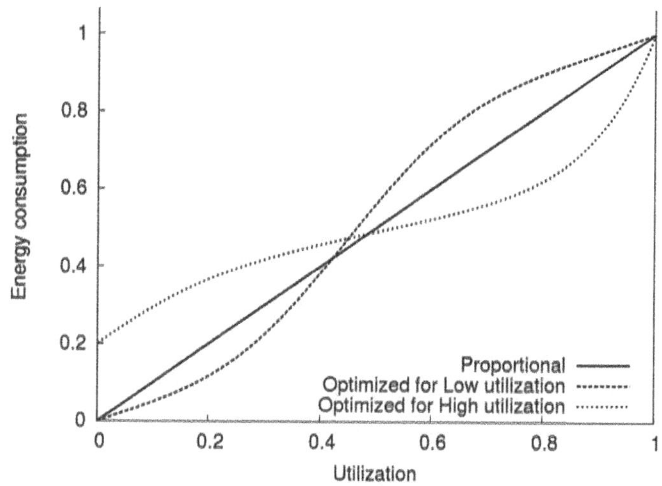

Figure 8.3. *The optimized carbon footprint for specific operational loads (Bianzino et al. 2012)*

8.4. Green wireless networks

A wireless network can be defined as: a network in which at least two terminals are able to communicate with each other using radio-electric signals.

In general, a wireless connection is used not only for convenience (no physical connection needed) and economic reasons (cost of the physical link is eliminated) but it is also used to overcome physical constraints (installation in dangerous areas, in areas that are difficult to access, within the human body, etc.).

Various sectors need to examine how their activities contribute to global warming and two significant sources of problems are:

– electromagnetic radiation: this is caused by radiation emitted by radio communication hardware, especially when they have a large range;

– CO_2 emissions: these are caused by the huge energy consumption of radio devices.

The telecommunications sector has integrated this need to reduce energy consumption. However, energy management becomes problematic in a wireless data link as the equipments require a power supply and must transmit information at a given transmission power.

This transmission power is simply the energy developed by the radio device to transmit its data and this is the cause for its power consumption, because if we wish to increase the energy developed (transmission power), we need to have a large source of energy (supply). Therefore, given the rise in equipment that is becoming increasingly miniaturized and/or difficult to recharge, it is essential to cut down the energy consumption and, thus, the energy costs of these links that are going to be deployed more and more widely.

Energy consumption models are as follows:

– putting forth the hypothesis that all mobile nodes in wireless networks are equipped with an IEEE 802.11 network card, whose data rate is 54 Mbit/sec, the energy required to transmit a packet, *p*, from node *i* is:

$$E_{tx}(p, n) = i * v * t_p \text{ J}$$

where i is the current (amperes), v is the voltage (volts) and t_p is the time taken to transmit the packet p (seconds) (Hedna 2011);

– based on the theory seen in Bloem *et al.* (2007), demonstrating that the longevity of a battery is inversely proportional to the transmission power, Grace *et al.* (2009) express energy consumption with respect to the battery's lifespan:

$$h_{Batt} = \frac{C_{Batt} * V_{Batt}}{P_{tx} + K} \text{ (h)}$$

where V_{Batt} is the voltage of the battery, P_{tx} is the transmission power and K is a constant that represents the power required to process signals.

8.4.1. *Energy efficiency in wireless networks*

The question of energy efficiency in wireless networks has been intensely studied. In reality, the lifespan of various pieces of wireless multisupply equipment has been flagged as problematic. In 2001, an important overview on the subject of network protocols on energy efficiency for wireless networks was presented by Jones *et al.* (2001) and this line of research has remained active ever since (Jiang *et al.* 2011).

The authors propose a method, in a project discussed in Hedna (2011), which could modify the Optimized Link State Routing Protocol (OLSR) routing protocol in order to promote a reduction in energy. This improvement is based on the interaction between the OLSR routing protocol and an enhancement of the MAC protocol, called Power-Aware Alternation (PAA), that is based on shutting down all network activity at a node during certain periods in order to conserve energy. During its period of inactivity, the messages at its destination will be taken up by one or more nodes, called "supporters". These supporters are chosen from a set of Multi-Point Relays (MPR) in the OLSR protocol, taking into account their energy level constraint. Thus, with this adaptation, the nodes with low energy are removed from the routing in order to conserve an adequate level of energy for mobile nodes.

Karray (2010) develops analytical models for spectrum efficiency and energy efficiency in networks using Orthogonal Frequency-Division Multiple Access. These models make it possible to study efficiency based on different parameters, such as the propagation exponent and the cell radius.

On the other hand, in Ouni *et al.* (2013), the authors implemented two algorithms for ordering and for managing "sleep mode" in Long-Term Evolution (LTE) advanced relay networks. These algorithms allow resources to be allocated to nodes and also manage their activities based on their states (transmission, reception or sleep) in order to reduce energy consumption. This heuristic makes it possible to calculate the ordering of the network nodes and their partial sleep mode using two different strategies based on the activation time of the nodes. The first strategy, called the LAFS (the Least Active, the First Served), starts by ordering the least active nodes in order to reduce the number of state changes and thus minimize the total "awakening cost". The second strategy is called MAFS (the Most Active, the First Served) and this starts by ordering the most active nodes in order to guarantee that these nodes can shift into partial sleep mode.

In the context of cellular systems, designers and network operators have considered decreasing power over the long term. However, with the increasing pressure to cut down on energy consumption and reducing greenhouse gas emissions, the concept of green cellular networks is gaining ground.

Most of the technologies used to minimize energy consumption in wireless systems try to optimize the transmission power of transmission–reception stations (Krief 2012).

8.4.2. *Controlling transmission power*

There is a problem with controlling power in *ad hoc* networks, related to the choice of the transmission power for each packet to be distributed. The problem is a complex one given that the choice of the power level influences multiple aspects of the functioning of the network:

– the transmission power level indicates the quality of the signal received by the recipient;

– it determines the range of a transmission;

– it provides an accurate measure of the magnitude of interference it will create for other recipients.

In conclusion, controlling transmission power in *ad hoc* networks is significant for at least three reasons (Narayanaswamy *et al.* 2002):

– it has an impact on the lifespan of the battery;

– it has an impact on network capacity in terms of the traffic transported;

– it may reduce radio-wave interference and, thus, amplify the spatial reuse of bandwidth.

According to Krunz *et al.* (2004), the transmission power determines the range over which the signal may be received harmoniously and, consequently, this is a critical parameter in determining the network performance (flow rate, time and energy consumption). There is in-depth work in literature on the selection of the "best" transmission range. In fact, Gupta and Kumar (2000) show that the greater the range of the nodes, the more power is required for transmission. Further, a node's range directly influences the zone of interference. Increasing the range of a node also increases the probability of interference, the collision rate and the packet loss rate, and it also reduces the transmission capacity of the nodes.

Controlling power consists of adapting the range and power of transmission by nodes, in order to ensure there is minimal energy consumption while also maintaining the network's connectivity. This requires finding an optimal transmission for the network nodes, which may or may not be common, making it possible to optimize the energy consumed during communications (Narayanaswamy *et al.* 2002; Liu *et al.* 2008).

The mechanism proposed in Ramanathan and Rosales-Hain (2000) makes it possible to combine a node's power until it has a limited number of neighbors. However, this does not always ensure network connectivity. Sometimes, nodes may find themselves in separate islands with very few direct neighbors. Thus, Ramanathan and Rosales-Hain (2000) presented two centralized algorithms "CONNECT" and "BICONN-AUGMENT" (Biconnectivity Augmentation), to minimize the maximum power used by a node while maintaining the network's connectivity or biconnectivity. The authors define a network as being biconnected if the loss of a single node does not interrupt the network. The CONNECT algorithm is a simple iterative algorithm that keeps associating components until only one remains. Initially, each node is its own component. Pairs of nodes are then chosen in non-decreasing order of their mutual distances. If the nodes are part of different components, then the transmission power for each is high, so as to be sufficient to reach the other node. This is continued until the network is connected. Moving from a connected network to a biconnected network is done through the BICONN-AUGMENT algorithm, which uses the same principle as CONNECT for the iterative construction of the biconnected network. However, this algorithm may additionally include a postprocessing phase to ensure the minimum distribution of transmission powers per node through the ending of redundant connections.

The objective of the COMPOW (Narayanaswamy *et al.* 2002) protocol is to modify the power of the nodes in accordance with a common value. This power level is the minimum level required for network connectivity. This protocol proves the importance of bidirectional links, as a direct destination can only respond to its source if its transmission power is at least equal to that of the source. As a result, putting in place a common power makes it possible to provide bidirectional links. This protocol is also used to increase the network's transmission capacity with the smallest possible energy level or smallest possible range while retaining network connectivity. These challenges pose the problem of providing excellent network coverage as well as controlling partitioning. It is, therefore, possible to say that COMPOW works well if the nodes are distributed homogeneously in space, even though a single distant node can force all other nodes to use greater power.

Finally, the role of green networking in wireless networks can provide conflict management between signal power and energy saving, i.e. need-based management of radio parameters, automatically and adaptively. Hence, there is need for cognitive radio (CR) services.

8.5. How CR contributes to green networking

CR or Smart Radio is a recent communication paradigm that has drawn the attention of mobile users, network operators and service providers.

The term CR is used to describe a system that is capable of detecting and recognizing the context in which it is being used, with the aim of dynamically and autonomously modifying its operating radio parameters and learning from the results of its actions and from the environment in which it is used. This system is, therefore, able to perceive and understand its environment while being able to adapt to it automatically, and involves a reprogrammable and reconfigurable class of radio (Stevenson *et al.* 2009).

8.5.1. *The principle behind CR*

CR is a technology that uses smart networks and terminals to:

– detect the user's communication needs based on their usage pattern;

– deploy the radio resources and wireless services that are best suited to these needs. In order to do this, CR must be endowed with three main abilities (Akyildiz *et al.* 2006):

- *awareness*: this is the aptitude for being aware of its environment. A CR terminal will combine its spatial environment, spectral environment and the behavior of its users to better understand resource provision and how to provide excellent service;

- *adaptation*: this is the ability to adapt either to its environment (spectral or technological) or to the user (needs and safety);

- *cognition*: this covers the various mental processes from perceptual analysis of the environment to motor commands (through memorization, reasoning, emotions, language, etc.). Thus, this description goes beyond the unique framework of animal or human cognition. CR is, therefore, a system that can perceive its own environment, analyze it, memorize it and act based on this information.

8.5.2. *The cognition cycle*

The cognitive component of CR architecture brings together temporal organization, inference flows and control states. This cycle arranges this component in an obvious manner.

Stimuli are introduced in CR as sensory interruptions sent to the cognition cycle (see Figure 8.4) for a response. A CR of this kind will then analyze the environment, orient itself, create plans, take a decision and then act (Ben Dhaou 2011).

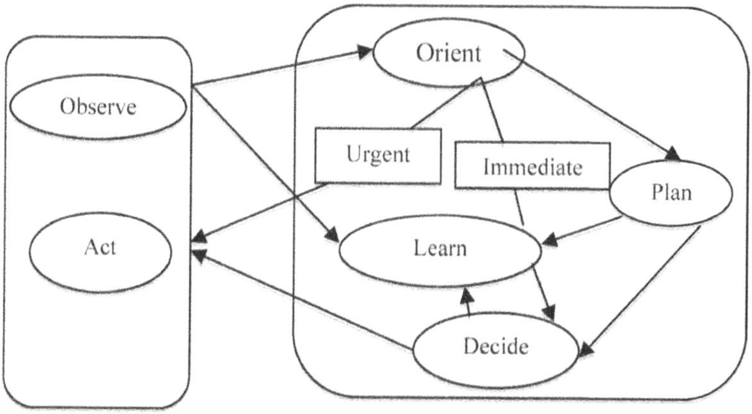

Figure 8.4. *The Mitola cognition cycle (Mitola 1998)*

8.5.2.1. *Observation phase (detection and perception)*

CR understands its environment through the analysis of the stimuli flow that it accesses. In the observation phase, the CR associates position, temperature, the light level recorded by the sensors, and so on, in order to deduce the context in which the communication is being made.

8.5.2.2. *Orientation phase*

The orientation phase determines the significance and scope of an observation phase by matching this with a known series of stimuli. This phase is carried out within data structures that can be compared to short-term memory in humans, which is used to carry out actions without the events necessarily being remembered in their entirety in the long term.

Stimuli are recognized when there is an exact match between a current stimulus and an earlier experience. The reaction may then be appropriate or erroneous.

Each stimulus is situated in a larger context, including other stimuli, that may be present, and the internal states, including time. The orientation phase often results in an action that is carried out immediately, like behavior during a "stimulus–response" reaction.

A power cut, for example, may result in an action to save data (the "immediate" path in the action phase depicted in Figure 8.4). A signal loss in a network may result in the re-allocation of resources, and this may be carried out through the path marked "urgent" in Figure 8.4.

8.5.2.3. *Planning phase*

A message entering a network will usually be processed by generating a plan (the normal path, in the planning phase). The plan must also include the reasoning plan over time.

8.5.2.4. *Decision-making phase*

In this phase, one action is selected from several possibilities. The radio may alert the user to an incoming message or report an interruption later, depending on the Quality of Information (QoI) levels that are defined for this phase.

8.5.2.5. *Action phase*

This phase triggers the chosen processes that use the selected effecters that access the external world or the CR's internal states. An action will be carried out on the environment, for instance, by making changes to the radio frequency level.

8.5.2.6. *Learning phase*

Learning depends on perception – one cannot exist without the other. You need observations, decisions and actions. The initial learning occurs through the observation phase, in which all sensory perceptions are compared to all previous experience, to take events into consideration and to keep in memory the time that has elapsed since the previous event.

8.5.3. *Green networking in CR networks*

Green networking may benefit from the larger paradigm CR offers. Haykin (2005) has described, in detail, CR equipment, such as a CR-electric transmitter, which will learn from the environment and adapt its internal states to the statistical variations of the input stimuli. In order to do this, the transmission parameters (e.g. the frequency band, the modulation mode and transmission power) must be adjusted in real time. In addition to this, we have cognitive abilities that can detect the available spectrum, as well as the reconfiguration ability to dynamically access this spectrum. The CR system increases the agility of the spectrum and energy efficiency (Mitola 1998).

There are very few sources in literature that focus on the health of the "green tech" user with respect to CR technology. Among the available sources, we wish to cite, in particular, the work carried out by Palicot (2009) and Michael *et al.* (2010) who have highlighted the harmful effects of using omnidirectional antennae in wireless terminals, which exposes the body, and, especially, the head, to high levels of electromagnetic radiation. The ideas and suggestions put forth in this document to reduce exposure to radiation include, notably, the use of directional antennae.

Indeed, there seems to be a need for the dynamic configuration of antennae, as well as of transmission powers and the other physical parameters associated with the signals emitted. However, it may be difficult to implement such adaptations, even though they are technically feasible with CR in the case of a mobile user and/or changing properties in transmission channels.

On the other hand, several research projects have been carried out with the aim of cutting down on energy consumption, especially during the detection step. Thus, in Maleki *et al.* (2011), the authors propose a combined sleep and censoring system to improve energy efficiency during spectrum detection in a cognitive sensor system.

They analyzed the performance of detection in this system by theoretically deducing the global probabilities for detection and false alarms. Their objective was to minimize the consumption of energy required for distributed sensing (given the constraints on the global probabilities of detection and false alarm) by defining an optimal sleep rate and censoring threshold.

In Elmachkour *et al.* (2014), the authors studied the conservation of energy and the performance of the system for allocating decentralized resources in CR networks based on competitive secondary users (SU). In their proposed approach, they apply the minority game to the most important phase of opportunistic access to the spectrum, i.e. the detection phases. The authors attempted to put cooperation in a non-cooperative environment in place, without exchanging information. They studied Nash's equilibrium solution for pure and totally mixed strategies, using distributed learning algorithms, which allow cognitive users to learn Nash's equilibrium. They concluded by providing the numerical results for the proposed approaches.

Furthermore, energy consumption may be significantly minimized by carrying out a direct action on communication infrastructure. Indeed, the energy consumption of pieces of equipment may be reduced by optimizing their operating mode or even by stopping them, which increases their lifespan and extends the recharge cycle. CR offers, in particular, the control of transmission power over each of the selected wavelengths. It is clear that the lower the transmission power the lower the energy consumed. However, if we decrease the transmission power, the rate and radius of coverage are impacted, which could be problematic. By choosing the right band of frequencies, CR can optimize these three parameters in order to improve energy efficiency.

In Grace *et al.* (2009), the deduction is specific: the authors demonstrate a direct relationship between the lifespan of the mobile terminal, the bandwidth and flow rate. Moreover, these researchers showed that the transmission power of a transmission–receptor station may have a negative impact on the energy consumption of terminals in a neighboring cell. Thus, as the transmission power of the access point increases, there is greater interference with non-recipients and neighboring cells must increase their transmission power in order to maintain the same ratio of signal to noise, which shortens their lifespan. By using different frequency bands in cells, CR is able to overcome this problem.

In order to minimize energy consumption, CR must take into account the relationship between the bandwidth, transmission power and capacity (flow rate). It is evident that if transmission power is reduced, the flow rate of communication will also be reduced. While it is possible to (slightly) compensate for this by increasing the bandwidth of the channel, it is not possible to do this over most of the frequency bands used. In addition, it is mentioned that transmission over low frequency channels can cover large distances, due to their resistance to attenuation and interference. In other words, if two distant nodes are not within each other's range using a given frequency, it is possible to use a lower frequency and the same transmission power to enable these two nodes to communicate. However, the flow rate of this communication will certainly be slower. In effect, by transmitting at a lower frequency, the bandwidth is also reduced (Krief 2012).

To conclude, it is possible to refine "green communications" by using CR, a technology that can dynamically adapt all the parameters involved in a transmission. The different parameters are represented in Figure 8.5.

The diagram shows that by choosing a low frequency, F1, and using low transmission power, a CR node, A, can communicate over equivalent distances to another node, B, which transmits at a higher power over frequency F2. This illustrates that CR, by dynamically selecting lower bands, can increase the lifespan of equipment and optimize the energy efficiency of networks. However, the gain in lifespan is obtained at the expense of the flow rate obtained. This constraint is not an obstacle when the lower frequencies chosen can provide a satisfactory flow rate for the communications. On the other hand, in some cases, the flow rate required for the communication may not be available when using low frequencies. Consequently, we will be forced to increase the frequency band and the transmission power.

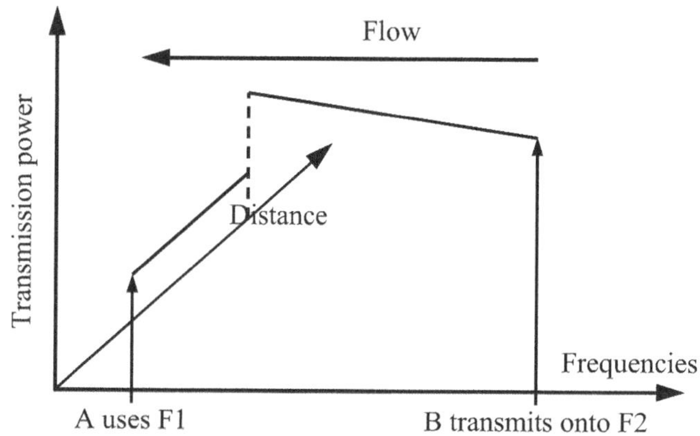

Figure 8.5. *Compromise: The bandwidth, transmission power, distance and flow rate (Krief 2012)*

8.6. Learning through reinforcement by taking into account energy efficiency during opportunistic access to the spectrum

Learning is defined as a concept that encompasses any method that makes it possible to construct a model of reality based on data.

Learning through reinforcement is a field of machine learning that makes it possible to resolve sequential decision-making problems when there is uncertainty.

Let us consider an agent (or a subject, actor, decision-maker) and an environment (or again, a system). The agent is confronted with a series of decisions to be taken. The environment is characterized by a state that changes over time in a random manner, based on the decisions taken by the agent. A decision can consist of choosing an action. This choice itself depends on the current state of the environment. Thus, the decision taken at each instant induces a state change which, in turn, influences the next decision to be taken. The interaction between the agent and the environment is characterized by a successive series of these events over time.

The term "learning through reinforcement" is used because at every instant the agent receives rewards (or gains) based on the state of the environment and the action that it has chosen. These rewards guide it in its future decision-making.

Thus, the agent's objective is to choose its actions so that is can maximize the rewards received. In general, the agent's goal is not to maximize the rewards received for every instance, but the long-term gains. The action that allows it to maximize an immediate reward given the state of the environment is, *a priori*, different from an action that will lead to a long-term benefit. Indeed, the reward received by the agent depends on both the chosen action and the state of the environment and, therefore, the agent must choose actions that may be less profitable in the short term in order to achieve a state where the reward is very large.

In this context, the team at "Centrale Supélec" (Modi *et al.* 2016) carried out a research project. The authors put forth a learning algorithm for a scenario with opportunistic spectrum access (OSA), modeled as a Markov multiarmed bandit (MAB) problem. The proposed algorithm chooses a channel for transmission which is optimal, not only in terms of the data flow rate but also in terms of quality, that is, the Signal to Noise Ratio (SNR). This allows SU to assign an appropriate weight to their desired criterion, such as the channel quality (which leads to reliable transmission with lower power) or the flow rate, by selecting two distinct exploration coefficients.

As the heart of the OSA problem is to learn which channels are best in terms of the chosen criteria, for instance, availability and quality. Recent projects have modeled the spectrum learning process based on the MAB framework (Jouini *et al.* 2010; Liu *et al.* 2013; Modi *et al.* 2015). The energy efficiency is as dependent on the propagation conditions in the channels as the spectral efficiency and, therefore, the system must adapt to the changing environment. Figure 8.6 illustrates the fact that certain channels are more available than others, due to the load of primary user traffic and that some of these channels offer better quality for wireless transmission in terms of the SNR, given the propagation conditions. Consequently, these channels could potentially support reliable transmission with lower transmission power.

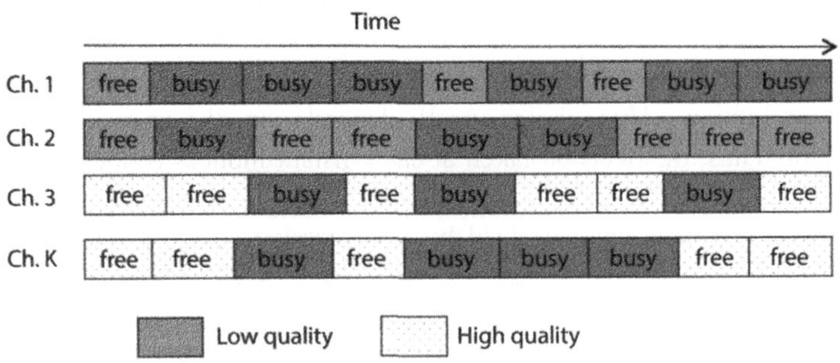

Figure 8.6. *State and condition of the channel for the SU. For a color version of this figure, see www.iste.co.uk/mbarek/service.zip*

Figure 8.7 presents the algorithm, which is specifically designed to solve the spectrum learning problem with a significant reduction in energy consumption for a secondary user. As with all flowcharts for learning through reinforcement, the proposed policy observes the availability of the channel through detection, which may be the output of the energy detector, in the case of CR.

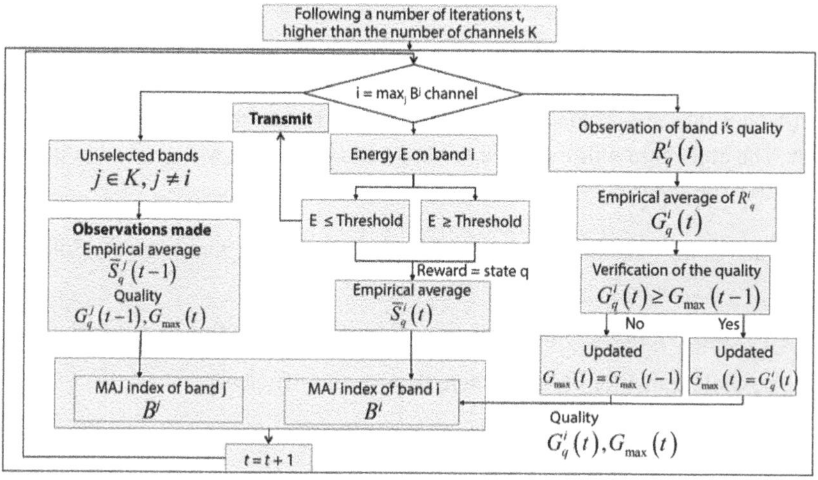

Figure 8.7. *New machine learning algorithm*

However, as we have already seen, reducing the transmission power leads to a reduction in flow rate and, thus, this problem still exists, in the case of a communication that requires a high flow rate.

We wish to refine this approach by rewarding or penalizing a channel based on the applicable criteria by implementing learning through reinforcement. The knowledge base that is built up will make it possible to choose a channel based on the user's requirements. Our suggestion is that a mechanism be added, in the CR context, to manage cases where the user communication requires a high flow rate. The learning through reinforcement in this case will be used to build up a knowledge base that can be used to store information that will help the CR terminal in future decision-making.

8.6.1. *Formulating the problem*

In a general case, in CR, the cognitive users communicate through a channel, taking into account only the channel availability. The study carried out by the

"Centrale Supélec" team proposed adding another constraint, that of transmission power, to minimize energy consumption. However, a problem arises when the transmission power is minimized, as the flow rate is also affected. Our approach aims to improve energy efficiency by selecting channels with low transmission power and a satisfactory flow rate for the communication by adopting learning through Reinforcement algorithm, "Q_learning".

Each channel must be considered as a state (S); moving from one channel to another is considered an action (A). Let us assume initially that all channels are available and that a minimum flow rate that is sufficient for the communication is known. The algorithm will require a reward matrix (R), and will produce a quantity matrix (Q).

For each state (channel), the matrix R contains a row that uses the following code:

– *1*: *(Penalty)* for impossible transitions (cases where the flow is not sufficient);

– *0*: possible transitions (cases where the flow is sufficient);

– *100*: *(Reward)* for the final state (among the possible transitions, the channel with the lowest power).

The matrix Q is now completed using the following algorithm:

For each cycle:

Select a random initial state.

While the final state is achieved, Do

Select one among all possible actions for the current state.

Using this possible action, consider moving to the next state.

Obtain the maximum value, Q, for this next state based on all possible actions.

Compute Q: *Q (state, action) = Q (state, action) + alpha * (R (state, action) + gamma * Max (next state, all actions) – Q (state, action))*.

Define the next state as the current state.

End while

End for

You may have noticed that two factors were used: alpha and gamma.

– Alpha is the rate of learning. The rate of learning determines the extent to which newly acquired information replaces older information. With a factor of 0, the agent will learn nothing, while for a factor of 1 the agent will only consider the most recent information. In our study, we used a constant learning rate of 0.8.

– Gamma is the reduction factor and determines the magnitude of future rewards. A factor of 0 makes the agent "myopic" and it only considers current rewards, while for a factor tending to 1 the agent can obtain higher rewards in the long term. In our study, we used a reduction factor of 0.2.

After having constructed matrix Q, we can predict a channel with low power and which will satisfy the communication as follows:

Begin from a given initial state, from this state, select the action with the highest Q value, in the case of an occupied channel; choose the second highest value.

8.6.2. *Comparison between CR and Q_learning enabled CR*

We wished to compare the results and energy efficiency in CR with and without Q_learning. To do this, we assumed there was a stable CR environment, two users who wished to communicate and required a large flow rate; we also assumed the existence of 100 channels, all of which are initially available for the users. We divided them with respect to the flow rate:

– *division 1*: 70 channels that will satisfy the communication and 30 unsatisfactory channels;

– *division 2*: 50 channels that will satisfy the communication and 50 unsatisfactory channels;

– *division 3*: 30 channels that will satisfy the communication and 70 unsatisfactory channels.

Using CR principles, the users detect the transmission power of each channel, as well as the proposed flow rate. In the general CR case, the user will hop from one channel to another in a random manner until they find a channel with a satisfactory flow rate. The algorithm that we propose makes it possible to evaluate channels based on their flow rate and the transmission power, in order to choose a channel that satisfies the communication, and to propose the lowest power based on the Q_learning principle, using the quantity matrix in this algorithm, which will constitute a knowledge base for future usage. Figure 8.8 represents a comparison with respect to energy consumption in the two cases over an average of 100 executions.

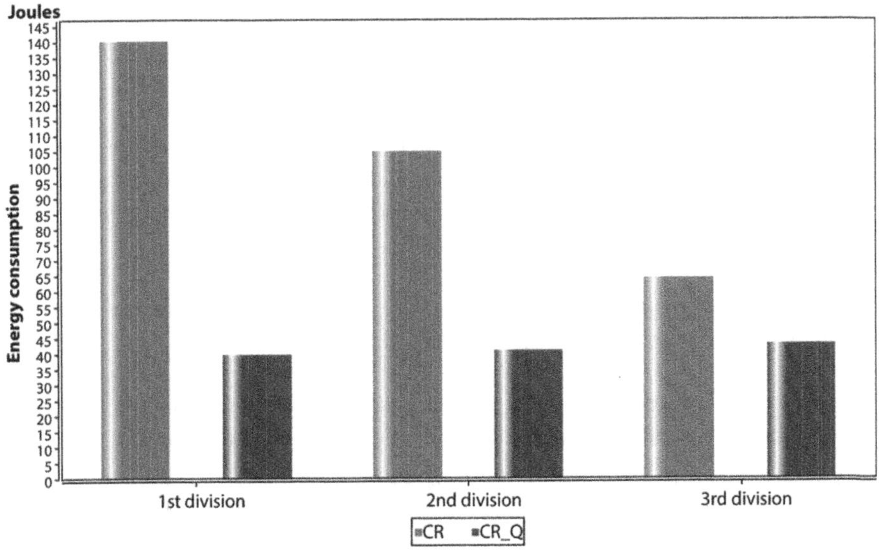

Figure 8.8. *CR versus CR with Q_learning. For a color version of this figure, see www.iste.co.uk/mbarek/service.zip*

There is a clear difference in energy consumption, due to the time taken by the user, in basic CR, to find a channel that fulfills their requirements, and also due to the fact that, in basic CR, a channel is randomly chosen, which increases the possibility of landing on a channel with high transmission power. On the other hand, when Q_learning is used, because of its learning, the user transmits over the appropriate channel at low transmission power.

8.7. Conclusion

We first introduced CR as a new paradigm in wireless communication, which came in with the aim of mitigating the lack of radio resources. CR was made possible by the dynamic allocation of frequencies. Further, because of its agility and ability to intelligently adapt communication parameters, CR allows us to minimize energy consumption and to move wireless networks toward "green networks".

We proposed a mechanism to make wireless communications more energy efficient. Our solutions involve using the properties of reconfiguration and adaptation that CR brings in, as well as gathering information on the frequency bands to be used. This gathering of information is possible because of the concept of learning through reinforcement and increases the lifespan of a network and reduces its carbon footprint.

ICT is subject to considerable constraints. Nonetheless, in order for green solutions to emerge, there must be collaboration between users. They must interact to coordinate and cooperate, for instance, to select lower flow rates. While CR is clearly a highly promising technology, it can only be optimally used if all actors involved make the required effort.

8.8. References

Akyildiz, I.F., Lee, W.-Y., Vuran, M.C., and Mohanty, S. (2006). Next generation/dynamic spectrum access/cognitive radio wireless networks: A survey. *Computer Networks*, 50(13), 2127–2159.

Allman, M., Christensen, K., Nordman, B., and Paxson, V. (2007). Enabling an energy-efficient future Internet through selectively connected endsystems. *Proceedings of the 6th ACM Workshop on Hot Topics in Networks (HotNets-VI)*. Atlanta.

Barroso, L.A. and Holzle, U. (2007). The case for energy-proportional computing. *IEEE Computer*, 40, 33–37.

Ben Dhaou, A. (2011). Allocation dynamique des bandes spectrales dans les réseaux sans-fil à radio cognitive. Thesis, Université du Québec, Montreal.

Bianzino, A.P., Chaudet, C., and Rossi, D. (2012). A survey of green networking research. *IEEE Communications Surveys & Tutorials*, 14(1), 3–20.

Bloem, M., Alpcan, T., and Basar, T. (2007). A Stackelberg game for power control and channel allocation in cognitive radio networks. *Workshop on Game Theory for Communication Networks 2007*. Nantes.

Chowdhury, N.M.K. and Boutaba, R. (2008). A survey of network virtualization. Report, University of Waterloo, Waterloo.

Christensen, K.J., Gunaratne, C., Nordman, B., and George, A.D. (2004). The next frontier for communications networks: Power management. *Computer Communications*, 27, 1758–1770.

Elmachkour, M., Daha, I., Sabir, E., Kobbane, A., and Ben-Othman, J. (2014). Green opportunistic access for cognitive radio networks: A minority game approach. *2014 IEEE International Conference on Communications (ICC)*. Sydney.

EPA (2007). Report to congress on server and data center energy efficiency public law 109-431. Report, U.S. Environmental Protection Agency, Washington.

ETNO, WWF (2007). Saving the Climate @ the Speed of Light: First roadmap for reduced CO_2 emissions in the EU and beyond. World Wildlife Fund and European Telecommunications Network Operators' Association, Brussels.

Grace, D., Chen, J., Jiang, T., and Mitchell, P.D. (2009). Using cognitive radio to deliver "green" communications. *4th International Conference on Cognitive Radio Oriented Wireless Networks and Communications*. 1–6, Hanover.

Gupta, P. and Kumar, P.R. (2000). The capacity of wireless networks. *IEEE Transactions on Information Theory*, 46(2), 388–404.

Haykin, S. (2005). Cognitive radio: Brain-empowered wireless communications. *IEEE Journal on Selected Areas in Communications*, 23(2), 201–220.

Hedna, S. (2011). Gestion de l'économie d'énergie dans les réseaux sans fil 802.11 Ad Hoc. PhD Thesis, University of Batna 2, Fesdis.

Jiang, C., Shi, Y., Hou, Y.T., and Kompella, S. (2011). On optimal throughput-energy curve for multi-hop wireless networks. *INFOCOM 2011. The 30th Conference on Computer Communications*. Shanghai.

Jones, C.E., Sivalingam, K.M., Agrawal, P., and Chen, J.C. (2001). A survey of energy efficient network protocols for wireless networks. *Wireless Networks*, 7(4), 343–358.

Jouini, W., Ernst, D., Moy, C., and Palicot, J. (2010). Upper confidence bound based decision making strategies and dynamic spectrum access. *International Conference on Communications, ICC'10*. Le Cap.

Karray, M.K. (2010). Spectral and energy efficiencies of OFDMA wireless cellular networks. *WirelessDays 2010*. 1–5, Venice.

Katz, R.H. (2009). Tech titans building boom. *IEEE Spectrum*, 46, 40–54.

Krief, F. (ed.) (2012). *Green Networking*. ISTE Ltd, London and John Wiley & Sons, New York.

Krunz, M., Muqattash, A., and Lee, S.J. (2004). Transmission power control in wireless ad hoc networks: Challenges, solutions, and open issues. *IEEE Network*, 18(5), 8–14.

Liu, Y., Guo, L., Ma, H., and Jiang, T. (2008). Energy efficient on-demand multipath routing protocol for multi-hop ad hoc networks. *IEEE 10th International Symposium on Spread Spectrum Techniques and Applications*. 572–576, Bologna.

Liu, H., Liu, K., and Zhao, Q. (2013). Learning in a changing world: Restless multi-armed bandit with unknown dynamics. *IEEE Transactions on Information Theory*, 59(3), 1902–1916.

Maleki, S., Pandharipande, A., and Leus, G. (2011). Energy-efficient distributed spectrum sensing for cognitive sensor networks. *IEEE Sensors*, 11(3), 565–573.

Michael, N., Moy, C., Vinod, A.P., and Palicot, J. (2010). Area–power trade-offs for flexible filtering in green radios. *Journal of Communications and Networks*, 12(2), 158–167.

Mitola, J. (1998). Cognitive radio. PhD Thesis, University of Stockholm, Stockholm.

Modi, N., Mary, P., and Moy, C. (2015). QoS driven channel selection algorithm for opportunistic spectrum access. *IEEE Globecom 2015 Workshop on Advances in Software Defined Radio Access Networks and Context-aware Cognitive Networks*. San Diego.

Modi, N., Moy, C., and Mary, P. (2016). Machine learning for opportunistic spectrum access with energy consumption constraint. *Journées scientifiques de l'URSI, Énergie et Radiosciences*. Rennes.

Nanda, S. and Chiueh, T.-C. (2005). A survey on virtualization technologies. Report, Department of Computer Science, SUNY, StonyBrook.

Narayanaswamy, S., Kawadia, V., Sreenivas, R.S., and Kumar, P.R. (2002). Power control in ad-hoc networks: Theory, architecture, algorithm and implementation of the COMPOW protocol. *European Wireless Conference*. Florence.

Nordman, B. and Christensen, K. (2005). Reducing the energy consumption of network devices. Tutorial, Universities of Berkeley and South Florida, San Francisco.

Ouni, A., Rivano, H., and Valois, F. (2013). Réduction de la consommation d'énergie des réseaux cellulaires hétérogènes. *15èmes Rencontres Francophones sur les Aspects Algorithmiques des Télécommunications*. 1–4, Pornic.

Palicot, J. (2009). Cognitive radio: An enabling technology for the green radio communications concept. *Proceedings of the 2009 International Conference on Wireless Communications and Mobile Computing: Connecting the World Wirelessly*. 489–494, Leipzig.

Qureshi, A., Weber, R., Balakrishnan, H., Guttag, J., and Maggs, B. (2009). Cutting the electric bill for internet-scale systems. *Proceedings of the ACM Conference on Applications, Technologies, Architectures, and Protocols for Computer Communications (SIGCOMM 2009)*. Barcelona.

Ramanathan, R. and Rosales-Hain, R. (2000). Topology control of multihop wireless networks using transmit power adjustment. *Proceedings of INFOCOM 2000*. 404–413, Tel Aviv.

Roth, K.W., Goldstein, F., and Kleinman, J. (2002). Energy consumption by office and telecommunications equipment in commercial buildings, volume I: Energy consumption baseline. Report, National Technical Information Service (NTIS), US Department of Commerce, Springfield.

Stevenson, C.R., Chouinard, G., Lei, Z., Hu, W., Shellhammer, S.J., and Caldwell, W. (2009). IEEE 802.22: The first cognitive radio wireless regional area network standard. *IEEE Communications Magazine*, 47(1), 130–138.

TC Group (2008). SMART 2020: Enabling the low carbon economy in the information age – Global e-Sustainability Initiative [Online]. Available at: http://www.gesi.org [Accessed March 2018].

Weiser, M., Welch, B., Demers, A., and Shenker, S. (1994). Scheduling for reduced CPU energy. *Proceedings of the 1st USENIX Conference on Operating Systems Design and Implementation (OSDI 1994)*. Monterey.

List of Authors

Hossam AFIFI
Telecom SudParis
Institut Polytechnique de Paris
Saclay
France

Hajer BARGAOUI
LIB
University of Bourgogne Franche-Comté
Dijon
France

Mohammed Salih BENDELLA
STIC
University of Abou Bekr Belkaïd
Tlemcen
Algeria

Badr BENMAMMAR
LTT
University of Abou Bekr Belkaïd
Tlemcen
Algeria

Chérifa BOUCETTA
LIPADE
University of Paris
France

Mohamed-Aymen CHALOUF
IRISA
University of Rennes 1
Lannion
France

Samira CHOUIKHI
ICD
University of Technology of Troyes
France

Benoît DARTIES
LIRMM
University of Montpellier
and
CNRS
Montpellier
France

Moez ESSEGHIR
ICD
University of Technology of Troyes
France

Ahmad KHALIL
LIB
University of Bourgogne Franche-Comté
Dijon
France

Nader MBAREK
LIB
University of Bourgogne Franche-Comté
Dijon
France

Leila MERGHEM-BOULAHIA
ICD
University of Technology of Troyes
France

Axel MOINET
LIB
University of Bourgogne Franche-Comté
Dijon
France

Hassine MOUNGLA
LIPADE
University of Paris
and
Telecom SudParis
Institut Polytechnique de Paris
Saclay
France

Olivier TOGNI
LIB
University of Bourgogne Franche-Comté
Dijon
France

Index

D

E, G

H, I

L, M

N

O, P

Q, R

S, T

U, W

Printed and bound by CPI Group (UK) Ltd, Croydon, CR0 4YY

13/03/2020

14961043-0002